A Simple Approach to College Algebra and Trigonometry

A Simple Approach to College Algebra and Trigonometry

Edward L. Green
Interboro Institute

Jerry Kornbluth
Interboro Institute

Australia · Canada · Mexico · Singapore · Spain · United Kingdom · United States

A Simple Approach to College Algebra and Trigonometry

Green and Kornbluth

Editor:
Jason Fremder

Marketing Coordinators:
Lindsay Annett and Sara Mercurio

Production/Manufacturing Supervisor:
Donna M. Brown

Project Coordinator:
Jennifer Flinchpaugh

Pre-Media Services Supervisor:
Dan Plofchan

Sr. Pre-press Specialist:
Kim Fry

Rights and Permissions Specialist:
Kalina Hintz and Bahman Naraghi

Cover Image
Getty Images*

Cover Designer:
Phoenix Creative, LLC

Compositor:
Integra Software Services Pvt. Ltd.

Printer:
BR

Printed in the
United States of America
1 2 3 4 5 6 7 8 08 07 06 05

For more information, please contact Thomson Custom Solutions, 5191 Natorp Boulevard, Mason, OH 45040. Or you can visit our Internet site at www.thomsoncustom.com

The Adaptable Courseware Program consists of products and additions to existing Thomson products that are produced from camera-ready copy. Peer review, class testing, and accuracy are primarily the responsibility of the author(s).

A Simple Approach to College Algebra and Trigonometry - Green and Kornbluth – First Edition
ISBN 0-759-36020-0

Library of Congress Control Number:
2005908079

International Divisions List

Asia (Including India):
Thomson Learning
60 Albert Street, #15-01
Albert Complex
Singapore 189969
Tel 65 336-6411
Fax 65 336-7411

Australia/New Zealand:
Thomson Learning Australia
102 Dodds Street
Southbank, Victoria 3006
Australia

Latin America:
Thomson Learning
Seneca 53
Colonia Polano
11560 Mexico, D.F., Mexico
Tel (525) 281-2906
Fax (525) 281-2656

Canada:
Thomson Nelson
1120 Birchmount Road
Toronto, Ontario
Canada M1K 5G4
Tel (416) 752-9100
Fax (416) 752-8102

UK/Europe/Middle East/Africa:
Thomson Learning
High Holborn House
50-51 Bedford Row
London, WC1R 4L$
United Kingdom
Tel 44 (020) 7067-2500
Fax 44 (020) 7067-2600

Spain (Includes Portugal):
Thomson Paraninfo
Calle Magallanes 25
28015 Madrid
España
Tel 34 (0)91 446-3350
Fax 34 (0)91 445-6218

Table of Contents

Acknowledgements			vii
Unit I	**Algebra**		**1**
	Section 1	Introduction	2
	Section 2	Operations Involving Real Numbers	9
	Section 3	Order of Operations	13
	Section 4	Algebraic Expression	16
	Section 5	Addition and Subtraction of Algebraic Expressions	19
	Section 6	Exponent Rule for Multiplication and Monomial Multiplication	22
	Section 7	Multiplication of a Monomial by a Monomial	24
	Section 8	Multiplication of a Monomial by a Binomial	25
	Section 9	Multiplication of a Binomial by a Binomial	27
	Section 10	Multiplication of Polynomials	29
	Section 11	Evaluating Algebraic Expressions	31
	Section 12	Order of Operations in Algebra	33
	Section 13	More on the Rules of Exponents — Exponent Rule	37
	Section 14	Division Of Algebraic Expressions	38
Unit II			**40**
	Section 15	Solving Linear Equations	42
	Section 16	Solving Equations with Fractions	44
	Section 17	Solving Literal Equations	49
	Section 18	Solving Verbal Problems—Number Problems	52
	Section 19	Ratios and Proportions—Verbal Problems	55
	Section 20	Inequalities	59
	Section 21	Absolute Value	62
	Section 22	Relations and Functions	65
	Section 23	Graph of a Straight Line	68
	Section 24	Solving Simultaneous Equations – Algebraic Solutions – Graphic Solutions	77
	Section 25	Properties of Exponents	86
	Section 26	Factoring A Polynomial	89
	Section 27	Factors: The difference of Two Squares	91
	Section 28	Trinomial Factors	93
	Section 29	An Additional Factoring Problem	97

Section 30 Rational Expressions 101
Section 31 Complex Numbers 106
Section 32 Solving Quadratic Equations 112
Section 33 To Solve Quadratic Equations by using the Quadratic Formula 114
Section 34 Radical Equations 119

Unit III **120**

Section 35 Introduction 121
Section 36 Plane Angle (Degrees – Radians) 122
Section 37 Pythagorean Theorem 124
Section 38 Trigonometric Functions of an Acute Angle 127
Section 39 Trigonometric Functions of 30°, 45°, and 60° 130
Section 40 Values of Trigonometric Functions of Acute Angles other than 30°, 45° and 60° 132
Section 41 Using Table to Evaluate Trigonometric Functions 134
Section 42 To solve for the missing angles and sides of a right triangle 136
Section 43 Reducing Trigonometric Functions To Positive Acute Angles 140
Section 44 Functions of a Negative Angle 142
Section 45 Reference Angles 144
Section 46 Basic Relationship and Identities 147
Section 47 Trigonometric Functions Sum – Difference – Product 150
Section 48 Solving Triangles That Do Not Have a Right Angle — Oblique Triangles 152

ACKNOWLEDGEMENTS

This book is dedicated to the encouragement and support my family has given me through the years. During my childhood my mother, Fae Sarah Wigder Green and my brother, Dr. William Green, encouraged me to pursue my education. My father, Jack Green, worked long hours to support his family. Furthermore, my aunt, Rose Saldinger, Molly Wigder and Ida Wigder assisted me to obtain my goals. My brother-in-law, Jose Chavin, gave me technical assistance.

I am very fortunate in that my beautiful wife, Zelma, always encouraged me to pursue my career. While I was working late at night, she graduated from Brooklyn College with her Masters Degree in Spanish literature while taking excellent care of our three wonderful sons, Elliot, Philip Steven and Seth Andrew. My oldest son, Elliot, and his wife Holly have two daughters, Isabella Morgan, and Emma Reese. My youngest son, Seth and his wife Edit also have one daughter, Sivon Michelle and a son Aoron Jack. I am extremely proud of my daughter-in-laws and grandchildren.

Professor Edward L. Green

I would like to thank my wonderful wife Jane for all her support and understanding throughout the writing of this book. Special thanks to my children, Brett, Scott, Nancy and Missy. The light of my life, Austin Cole, Hannah and Alexa, motivated me to write a book that they might use in the future.

Dr. Jerry Kornbluth
Professor Emeritus

With great joy and gratitude to my wife, Bessie, who remained always a true companion during both the good and the bad. Special thanks must also go to my two daughters Lynn and Lindsey. All three have been instrumental in making it possible for me to complete present project in a record time.

Professor Paul C. Tsang

Dedicated to the loving memory of my dad, my mom and my son. They have served as my inspiration, my spiritual lighthouse, and my intellectual guide in my journey through life.

Professor Nestor Baylan

UNIT

one

ALGEBRA

Section 1: Introduction 2

Section 2: Operations Involving Real Numbers 9

Section 3: Order of Operations 13

Section 4: Algebraic Expression 16

Section 5: Addition and Subtraction of Algebraic Expressions 19

Section 6: Exponent Rule for Multiplication and Monomial Multiplication 22

Section 7: Multiplication of a Monomial by a Monomial 24

Section 8: Multiplication of a Monomial by a Binomial 25

Section 9: Multiplication of a Binomial by a Binomial 27

Section 10: Multiplication of Polynomials 29

Section 11: Evaluating Algebraic Expressions 31

Section 12: Order of Operations in Algebra 33

Section 13: More on the Rules of Exponents — Exponent Rule 37

Section 14: Division Of Algebraic Expressions 38

SECTION 1

INTRODUCTION

Algebra is a course that cannot be learned by observation or by kicking your chair back and watching. To learn algebra you must be an active participant. In other words, you have to get on the playing field. You must read the text book, pay attention in class, and most important, you must do your homework. The more exercise you do, the better you will be. Do the problem sets at the end of the chapter.

This text was written with you in mind. Short, clear sentences are used and many examples are given to illustrate specific points. It is our goal that you come to realize that algebra is not just another math course that you are required to take, but a course that offers a wealth of useful information and applications. You may wish to form a study group with other students in your class. Many students find that working in small groups provides an excellent way to learn the material. You may be thinking to yourself, "I hate math," or "I wish I did not have to take this class." You may have picked up or heard of the term, "Math Anxiety," and feel you fit this category. In order to be successful in this class you have to change your attitude to a more positive one. As authors of this book as well as instructors for this type of course, we will help you achieve this new positive attitude.

There are two very important commitments that you must make to be successful in this course. They are attending class and doing your homework regularly. You need to practice what you have heard in class. It is through doing homework that you truly learn the material. Ask questions in class about homework problems you don't understand. You should not feel satisfied until you understand all the concepts needed to successfully work each and every assigned problem.You should plan to attend every class. There is a positive relationship between grades and attending class. Every time you miss a class, you miss important information. There is a saying that for every missed class, it takes about three classes to make it up. If you need to miss a class, contact your instructor ahead of time, and get the homework and reading assignment.

When you are in class, take careful notes. Write both numbers and letters clearly, so that you can read them later. Be sure to get help as soon as you feel you need it. Do not wait.

Where should you seek help? There are a numbers of resources on campus. You should know your instructor's office hours, e-mail address, and you should not hesitate to seek help from your instructor when you need it Make sure you have read the assigned material and tried the homework before meeting with your instructor.

Always come prepared with specific questions to ask. The worst thing that you could do is "throw up your hands" and say, "I don't understand this, help". You must make an attempt. This goes back to having a positive attitude.

There are often other sources of help available. Many colleges have a mathematics lab or a mathematics learning center, where tutors are available. Ask you instructor early in the semester if tutoring is available. Find out how to arrange for tutorial help and work with a tutor as needed.

Remember, follow these hints and not only will you be successful in math but you will also find math to be fun.

The Real Number System

During our study of Algebra we will only work with Real Numbers. I know that you have heard of and have seen this term many times.

The Real Numbers

- A Set if a collection of objects. Sets that are parts of other sets are called subsets.
- Natural Numbers: (1,2,3. . ..) These are sets of numbers used for counting.
- Whole Numbers: (0,1,2,3. . ..) This is the set of natural numbers with 0 included.
- Integers: Integers are whole numbers, not fractions.
- What are Real numbers? They are a set of all rational numbers and irrational numbers. All these numbers can be represented on a number line.
- Rational Numbers: Any number that can be represented as a single fraction with a non-zero divisor.
- Irrational Numbers: They are real numbers that are not rational:

$$\sqrt{3} \text{ or } \sqrt{5}.$$

- The Number Line: Is used to represent all positive numbers and all negative numbers.

Negative ◀ Positive ▶

$-4 \quad -3 \quad -2 \quad -1 \quad 0 \quad +1 \quad +2 \quad +3 \quad +4$

When representing a number on the number line, the distance of that number from 0 is called the absolute value. Thus, the absolute value of +2 is 2, and the absolute value of −2 is 2. The absolute value of a number is its distance from 0 on the number line. The number is represented by the symbol || from 0 on the number line. The number is represented by the symbol ||.

Let Us Explore These Real Numbers

Integer: An integer is a whole number, which can be either negative(-) or positive (+) and zero (0).

Examples: 5, −2, 300, 6, 0.

Rational Number: A fraction $\frac{a}{b}$, where both a and b are integers and $b \neq 0$.

Example: $\frac{3}{4}, -\frac{5}{2}, -\frac{10}{3}, \frac{12}{5}, \frac{2}{3}$

What about the number 6? We can express 6 as 6/1, So what does this mean? The number 6 is a rational number because we can write it as a fraction: 6/1. All Integers are reational numbers, but not all rational numbers are integers.

What about 6/0? In the definition of a rational number, we can see that there can never be a 0 in the denominator. So 6/0 is what we call undedfined. It is correct, however, to have a 0 in the numerator. For example 0/6 = 0. What about 0/0? I bet you would like to say that the answer is 1. You know that you can never have a 0 in the denominator. So 0/0 is undefined.

Fractions can also be expressed as a decimal. For example:

EX. 1 $\frac{2}{5} = 0.4$ EX. 2 $\frac{3}{4} = 0.75$ EX. 3 $\frac{1}{4} = 0.25$

These decimals are called terminating decimals. These are also rational.

Now, What about

$$\frac{1}{3} = .3333\ldots = .\bar{3}$$

$$\frac{1}{9} = .1111\ldots = .\bar{1}$$

$$\frac{2}{3} = .6666\ldots = .\bar{6}$$

These decimals are called repeating decimals. Decimals can repeat in 1 digit, 2 digits, or as many digits as possible. These decimals are also called rational numbers. So is that it? Are there any other numbers in the Real Number System? You bet.

Irrational Number: Most books would probably define this as a number that is not rational. Let me do a little better job of defining an irrational number. We can say that an irrational number is a decimal that does not repeat and does not terminate. Is this a little better? The classic irrational number is Π It is 3.14...... In order to terminate the decimal, we must round it off. So Π or any combination of Π such as 2Π or Π/3 is a non-repeating, non-terminating decimal. The easiest way to identify any other irrational number is that they are square roots of all non-perfect squares.

Example: $\sqrt{4} = 2$, $\sqrt{9} = 3$, $\sqrt{25} = 5$,

These numbers are perfect squares that happen to be integers and rational numbers.

How about these numbers?

$$\sqrt{2} = 1.41421356237\ldots.$$

$$\sqrt{3} = 1.73205080756\ldots.$$

$$\sqrt{8} = 2.82842712474\ldots.$$

So you see these numbers never end and never repeat themselves. These are called irrational numbers.

Let Us Summarize

The Real Number System

Rational Numbers

Irrational Numbers

Integers

Properties of Real Numbers

The properties of real numbers are defined extensively in other college mathematics courses. In this text we will explain through examples those properties of real numbers that are most important in the study of algebra; namely, the commutative, associative, distributive, inverse, and identity laws.

The Commutative Law

This property of real numbers states that the sum and product of any two real numbers is the same, regardless of the way in which you add or multiply.

$$6 + 3 = 3 + 6$$

$$3 \times 4 = 4 \times 3$$

$$0.6 + 0.4 = 0.4 + 0.6$$

$$0.6 \times 0.4 = 0.4 \times 0.6$$

So here is the rule:

$$a + b = b + a$$

$$a \times b = b \times a$$

$$ab = ba$$

The key word is order. If you want to change the order without changing the value, you can then use the Commutative Law. Also, what about: $5\ (2A + B) = 5\ (B + 2A) = (2A + B)\ 5$? Notice, all we did here is change the order. We never changed the parentheses.

The Associative Law

The associative property of real numbers states that the sum or product of any three real numbers is the same, regardless of the order in which you add or multiply.

$$6 + (4 + 3) = (6 + 4) + 3$$

$$6 \times (4 \times 3) = (6 \times 4) \times 3$$

So there is the rule:

$$a + (b + c) = (a + b) + c$$

$$a \times (b \times c) = (a \times b) \times c$$

$$a(bc) = (ab)c$$

The key word is regroup. The easiest way to remember the Associative Law is to keep in mind three things:

1. Regroup; change parentheses.
2. Keep the same order.
3. Associate only over the same operation.

For example, $5A + (3B + 2A)$. Our goal as you will see later is to put like things together. That means I would like to get the 5A with the 2A. Here is why these rules are important. Let us follow:

$$5A + (3B + 2A) = 5A + (2A + 3B)$$

What did I do here? If you remember, all I did was change the order. We call that the Commutative Law.

Then I want to regroup.

$$(5A + 2A) + 3B = 7A + 3B$$

We will explore this later.

The Distributive Law

This property actually involves the two operations of multiplication and addition or multiplication and subtraction.

We might say that multiplication is "distributed" over a sum, or multiplication is "distributed" over a subtraction (difference).

$$2(4 + 2) = 2(4) + 2(2) = 8 + 4 = 12$$

$$2(4 - 2) = 2(4) - 2(2) = 8 - 4 = 4$$

Why distribute if I know that:

$$2(4 + 2) = 2(6) = 12$$

$$2(4 - 2) = 2(2) = 4$$

This seems easier and you happen to be right. However what about this?

$$5(2A + 3B) = 5(2A) + 5(3B) = 10A + 15B$$

$$5(2A - 3B) = 5(2A) - 5(3B) = 10A - 15B$$

As you can see above, you cannot add or subtract inside the parentheses because you do not have like terms.

So you must use the distributive law. So the key word in the distributive property is parentheses. When you see a parentheses you want to multiply and use the Distributive Law.

Identity Law

There is an identity property for addition and for multiplication. The number 0 (zero) is called the identity of addition (or subtraction), since adding (or subtracting) zero does not change the identity of a number.

Examples: $5 + 0 = 5$, and $5 - 0 = 5$.

The number 1 (one) is the identity for multiplication (or division), since multiplying (or dividing) by 1 (one) does not change the identity of a number.

Examples: $5\ (1) = 5$, and $5/1 = 5$.

Remember that the number 1 (one) may also be expressed as 6/6, 3/3, 1.5/1.5. In general, $a/a = 1$.

So what are the key words here? The key words are change the look. Multiplying or dividing by 1 (one), or adding or subtracting by 0 (zero) does not change the value.

Inverse Law

This property is used to solve an equation. There is the inverse law for addition and for multiplication.

Additive Inverses are numbers whose sum is 0 (zero).

Example:
$$a + (-a) = 0$$
$$5 + (-5) = 0$$
$$-5 + 5 = 0$$

Multiplicative Inverses

Reciprocals are numbers whose product is 1 (one).

Examples:
$$a \bullet \frac{1}{a} = 1$$
$$5 \bullet \frac{1}{5} = 1$$

So again, the key here is for solving an equation. We will explore this later.

Let Us Summarize

Key Word	
1) Commutative Law	Order
$a + b = b + a$	
$a(b) = b(a)$	
2) Associative Law	Regroup
$a + (b + c) = (a + b) + c$	
$a(bc) = (ab)c$	
3) Distributive Law	Remove Parentheses
$a(b + c) = ab + ac$	
$a(b - c) = ab - ac$	
4) Identity Law	Change the Look
$a + 0 = a$	
$a - 0 = a$	
$a \bullet 1 = a$	
$a \div 1 = a$	
5) Inverse Law	Solve an Equation
$a + (-a) = 0$	
$a \times \frac{1}{a} = 1$	

Problem Set #1

I. Fill out the following Real Number Table:

Number	Real	Integer	Rational	Irrational	None of these
Example:					
5.2	√		√		
1. $\frac{3}{5}$					
2. 6.285					
3. $\sqrt{18}$					
4. 2Π					
5. $2\sqrt{25}$					
6. 8%					
7. 200%					
8. 5/0					
9. $-\sqrt{49}$					
10. $\sqrt{-4}$					

II. State the property of real numbers (commutative, associative, distributive, identity, inverse) illustrated by each of the following:

1. $20 + 30 = 30 + 20$
2. $6 + (3 + 2) = (6 + 3) + 2$
3. $3(5 \times 4) = 3(5) + 3(4)$
4. $5 + 0 = 0 + 5$
5. $2A + 3B = 3B + 2A$
6. $5(0) = 0(5)$
7. $6 \bullet 1 = 1 \bullet 6$
8. $4 + 0 = 4$
9. $4 \bullet 1 = 4$
10. $\frac{1}{5} \times 5 = 1$
11. $4 + (-4) = 0$
12. $5(4 \times 2) = 5(2 \times 4)$

SECTION 2

Operations Involving Real Numbers

What we want to show you is how to simplify a numerical expression. What is a numerical expression?

It is an expression that includes numbers and operations. Where do we make our errors? The key word here is Signs. Let us look at some easy rules.

Rules for Signed Numbers: Addition, Subtraction, Multiplication, and Division

Addition of Signed Numbers

1. When adding numbers with the same signs, keep the sign and added the numbers.
2. When adding numbers with the opposite signs, keep the sign of the larger number and take the difference.

Example 1: $(-3) + (-5)$

Answer: -8

Example 2: $(-6) + (8)$

Answer: 2

Example 3: $-4 - 7 = -11$

Example 4: $6 + 2 = 8$

Hint: You add when the signs are the same. Look of examples 3 and 4.

Subtraction of Signed Numbers

When subtracting numbers, change the sign of the subtrahend (the number after the subtraction sign) to the opposite sign and then follow rules for addition.

Example 1: $6 - (-5)$

Solution: The opposite of -5 is 5

$-(-5) = 5$

$6 - (-5) = -6 + 5 = 11$

Answer: 11

Example 2: $-6 - (-5)$

Solution: The opposite of -5 is 5

$-(-5) = 5$

$6 - (-5) = -6 + 5 = -1$

Answer: -1

Example 3: $-2 + 8 = 6$

Answer: 6

Example 4: $6 - 10 = -4$

Answer: -4

Hint: You subtract when the signs are the opposite. You take the smaller number from the larger number and keep the sign of the larger number. Look at examples 3 and 4.

Part A

Fill in the correct answer in the space provided:

1. $(-3) + (-4)$ Answer ________
2. $-2 + (7)$ Answer ________
3. $(-9) + (12)$ Answer ________
4. $(7) + (-4)$ Answer ________
5. $-3 + (-8)$ Answer ________
6. $6 + (-7)$ Answer ________
7. $-8 + 4 - 7 + 6 - 3 + 8$ Answer ________
8. $-2 - 7 + 6 - 3 + 2 + 1$ Answer ________

Part B

Fill in the correct answer in the space provided:

1. $(-2) - (-6)$ Answer ________
2. $(-3) - (-7)$ Answer ________
3. $-2 - (-7)$ Answer ________
4. $4 - (-7)$ Answer ________
5. $6 - (-3)$ Answer ________
6. $-6 - (-2)$ Answer ________
7. $9 - (-5)$ Answer ________

8. $-8 - (-6)$ Answer ________

9. $-1 - (-10)$ Answer ________

10. $-3 - (-4)$ Answer ________

11. $(-2) - (-7)$ Answer ________

12. $3 - (-8)$ Answer ________

Multiplication And Division Of Signed Numbers

1. When mutiplying or dividing two numbers with the same sign, the result is positive.
2. When mutiplying or dividing two numbers with the opposite signs, the result is negative.
3. The product of an odd number of negative numbers is negative.
4. The product of an even number of negative numbers is positive.

For Example:

$6 \times 7 = 7 \times 6 = 42$
$(3 \times 4) \times 5 = 3 \times (4 \times 5) = 60$

Example 1: $3(-7)$

Solution: $3(-7) = -21$
Answer: -21

Example 2: $(-3)(-4)$

Solution: $(-3)(-4) = 12$
Answer: 12

Example 3: $(-2)(-7)(-1)(-5)(-6)$

Solution: Multiplication. An odd number of negative signs–the answer is negative
Answer: -420

Example 4: $(-3)(-4)(-1)(-5)(-2)(-6)$

Solution: Multiplication. An even number of negative signs–the answer is positive
Answer: 720

Division facts:

1. $a \div a = 1$ A number divided by itself equals 1.
2. $a \div 1 = a$ A number divided by 1 equals the number.
3. $0 \div a = 0$ Zero divided by a number equals zero.
4. $a \div 0 = \infty$ Undecided. A number divided by zero is undecided – not defined.

Example 1: $-6 \div 2$
Numerator and Denominator: Opposite signs in

Solution: $-6 \div 2$ Division. The result is negative (no sign by the number
Answer: -3 2 indicates it is positive.)

Example 2: $-9 \div -3$

Numerator and Denominator: Same signs in division

Solution: $-9 \div -3$ means the result is positive.

Answer: 3

Hint: When you divide or multiply and the signs are the same, the answer is positive. When you divide or multiply and the signs are opposite, then the answer is negative.

Fill in the correct answer in the space provided:

1. $(3)(-4)$ Answer ________
2. $7(-5)$ Answer ________
3. $-6(-2)$ Answer ________
4. $(-3)(5)$ Answer ________
5. $(-1)(-2)(-4)(-1)(-7)$ Answer ________
6. $(-3)(-1)(-4)(-1)(-5)(-1)$ Answer ________
7. $\frac{-63}{-7}$ Answer ________
8. $\frac{-15}{5}$ Answer ________
9. $\frac{32}{-8}$ Answer ________
10. $\frac{-9}{-3}$ Answer ________
11. $\frac{-14}{2}$ Answer ________
12. $\frac{63}{-7}$ Answer ________

Now that we all know how to add, subtract, multiply and divide, let us put it all together.

Do you think that you can do this? I think the answer is 8 or 11. Should we flip a coin? I hope not. So how do we do this? Do we work left to right or right to left, or may be inside out. Well, let us not guess or flip coins. What we need to look at is what we call the order of operations.

SECTION 3

ORDER OF OPERATIONS

RULES FOR ORDER OF OPERATIONS

1. Performing indicated operations inside the parentheses.
2. Evaluate the roots and the powers/exponent.
3. Multiply or divide from left to right.
4. If necessary, add or subtract.

- **Squaring a Number**: A number multiplied by itself
- **Power**: It is the number of times a basis is used as a factor
- **Exponent**: It is the same as powers
- **Base**: The number that serve as a starting point

Base Power

1. 3	4	3 : base	4: power/exponent
2. 6	7	6 : base	7: power/exponent

For Example:

1. a	3	$= a \cdot a \cdot a$	A dot between two variables or
2. 2	3	$= 2 \cdot 2 \cdot 2 = 8$	numerals indicates multiplication.
3. a	5	$= a \cdot a \cdot a \cdot a \cdot a$	
4. 3	5	$= 3 \cdot 3 \cdot 3 \cdot 3 \cdot 3 = 243$	

Example 1: $3(-2 + 4)$

Solution: $3(2)$

Answer: 6

Example 2: 3 − 2(4 + 5)2

Solution: 3 − 2(9)2
3 − 2(81)
3 − 162
−159

Answer: −159

Example 3: 4(−5)2 + 3(−7)

Solution: 4(25) + 3(−7)
100 − 21
79

Answer: 79

PROBLEM SET #2

1. −32
2. (−3)2
3. 6(7 + 1)
4. 8(4 − 2)
5. −2(8 − 4)
6. 6(3 − 7)
7. 2 − 3(7− 4)
8. 6 + 2(8 + 4)
9. 3(8 + 1) + 4
10. 7(4 + 3) − 3(5 + 9)
11. 4(6)2 2
12. −3(4 + 5)2
13. −3 + (4 + 5)2
14. 6 − 7(5)2
15. −3(−4)2
16. −6(−3 −4)
17. 3 + 2(4 + 5)2
18. (−3)2 + 4(−5)
19. (−4 −5)2
20. (−2)2 −4(−3)2
21. −7(4 − 2)2
22. −3(4 + 7)2

23. $-3 + 2(4 + 7)2$

24. $-4(9 + 1)2 - 7$

25. $7 - 8(6)2$

26. $(-4)2 + 3(-6)$

27. $(-2 -3)2$

28. $4 + 3(5 + 6)2$

29. $-2(-4 -5)2$

30. $5(7)2$

31. $6 - 2(-3)2$

32. $-3(-4)2 - 5(-6)2$

SECTION 4

ALGEBRAIC EXPRESSION

An algebraic expression is a statement that contains any or all of the following:

a. Literal part(s)

b. Numeric part(s)

c. Signs indicating operations

d. Grouping symbols

The literal part of an algebraic expression may change in value; thus it is called a variable.
The numeric part remains unchanged; (i.e. 3 is always 3); thus is it called a constant.

Example: The term 3x3 contains the variable x and constant 3 and 2. This single termed expression is called a monomial.

Example: The expression 3x2 – 4 contains the variable x, and the constant 2 (the exponent), 3 and 4. It also contains the operational sign. The two-termed expression is called a binomial.

Example: The expression 3x4 –(x3 +2) contains the variable x, the constant 3 and 2, the exponent, which is the constant 4, the operational sign, –, and the grouping sign symbol,().

Remembering that subtraction is the inverse of addition and (–x3 –2) is the additive inverse of x3 + 2, we may write: 3x4 – (x3 +2) as 3x4 – x3 – x3 – 2.
This three-termed expression is called a trinomial.

INTERPRETING ALGEBRAIC EXPRESSIONS

Variable

A symbol that stands for a number.

Example 1: Suppose that “b” represents a number

a. 2 less than a number

b. 5 more than a number

c. 3 times a number

d. the number divided by 4

Solution:

a. Less than represents −, b − 2

b. More than represents +, b + 5 or 5 + b

c. 3 times a number, 3 • b or 3b

d. The number divided by 4, b ÷ 4 or b/4.

Example 2: Find the value in cents of nickels and dimes.

Solution: One nickel is valued at 5 cents.
n nickel = 5 x n = 5 n
One dime is valued at 10 cents.
d dimes − 10 x d = 10d
n nickels and d dimes
5n + 10d

Answer: 5n + 10d

Interpreting each of the following algebraic expressions in terms of "x":

Fill in the correct answer in the space provided.

1. 5 more than a number. Answer
2. a number minus 7 Answer
3. a number decreased by 6 Answer
4. a number plus 5 Answer
5. 7 times a number. Answer
6. 4 less than a number Answer
7. a number increased by 14 Answer
8. 4 more than one-half of a number Answer
9. 3 more than 4 times a number Answer
10. 6 less than 3 times a number. Answer

Interpret each of the following algebraic expressions. Fill in the correct answer in the space provided.

Value of money

1. One nickel = 5 cents
"n" nickels = 5n
2. One dime = 10 cents
"n" dimes = 10d
3. One quarter = 25 cents
"n" quarters = 25q

1. Find the value in cents of "n" nickels and "q" quarters.
Answer

2. Find the value in cents of "d" dimes and "q" quarters.
Answer

3. Find the value in cents of “d” dimes and “n” nickels.
   Answer

4. Find the value in cents of “d” dimes and “n” nickels.
   Answer

5. Find the value in cents of “q” quarters, “n” nickels, and
   Answer
   “d” dimes.

Addition and Subtraction of Algebraic Expressions

SECTION 5

Rules:
Additions and subtraction of Algebraic Expressions
Add or Subtract coefficients (numbers in front) of like terms.

Like Terms

a. Same variables

b. Same exponents for the same variables.

Example 1: Which are like terms
a. $6x^2, 3x^2$ b. $4a^2, 6a^3$

Solution:
a.) Like terms: Both terms have the same variable and the same exponent:
$6x^2, 3x^2$
b.) Unlike terms: Both terms have the same variables. However, they have different exponents:
$4a^2, 6a^3$

Example 2: $(3a + 4b) + (5a - 7b)$

Solution: Line up like terms and add in columns
$3a + 4b$
$5a - 7b$
$8a - 3b$

Example 3: Find the sum of $3x^2 - 7x$ and $2 - 5x$

Solution: Line up like terms
$3x^2 - 7x$
$- 5x + 2$
$3x^2 - 12x + 2$

Example 4: Subtract $6x^2 - 9$ from $3x^2 - 7$

Solution: Line up like terms

1) $3x^2 - 7$
   $6x^2 - 9$
2) change the subtrahend(bottom number to the opposite) and then follow rules of addition.
   $3x^2 - 7$
   $-6x^2 + 9$
   $-3x^2 + 2$

Let Us Look At Another Approach When We Combine Like Terms

Back To Example 3

Example 3: Find the sum of $3x^2 - 7x$ and $2 - 5x$

Let us rewrite this statement: $(3x^2 - 7x) + (2 - 5x)$ Since an imaginary 1(one) is in front of both parentheses, we get $1(3x^2 - 7x) + 1(2 - 5x)$, we can use distributive law. This will give us $3x^2 - 7x + 2 - 5x$. Then combine like terms. The only like terms that are alike are the $-7x$ and $-5x$. Since the signs are the same we keep the sign and add. Now we get $-7x - 5x = -12x$. The final answer is : $3x^2 - 12x + 2$

Back To Example 4

Example 4: Subtract $6x^2 - 9$ from $3x^2 - 7$

Let us rewrite this statement: $(3x^2 - 7) - (6x^2 - 9)$. The from part always goes first. In other words, you will get from – ( ). Since there is an imaginary 1 (one) in front of both parentheses, you then get $1(3x^2 - 7) - 1(6x^2 - 9)$. Let us use the distributive law again: $3x^2 - 7 - 6x^2 + 9$. Now combine like terms: $3x^2 - 6x^2$ and $-7 + 9$.

Remember - Same signs you add and opposite signs you subtract.
$3x^2 - 6x^2 = -3x^2$
$-7 + 9 = 2$

Your Answer is = $-3x^2 + 2$

Fill in the correct answer in the space provided:

1. Add $3x + 7y$ and $6x + 8y$ Answer ________
2. Add $3a - 7b$ and $4a - 2b$ Answer ________
3. Add $6x^2 + 5x$ and $8x^3 - 7x$ Answer ________
4. Add $7b - 2$ and $3b + 7$ Answer ________
5. Find the sum of $4x^2 - 7x + 6$ and $-3x + 2$ Answer ________
6. Find the sum of $7y^2 + 3y$ and $3y^2 + 7y - 8$ Answer ________
7. Add $3a^2 - 7a + 6$ and $-7a^2 + 4a - 8$ Answer ________
8. Subtract $-3x + 6$ from $x - 8$ Answer ________
9. Subtract $-7x^2 + 8x$ from $x^2 - 8x$ Answer ________
10. Subtract $-3x^2 - 7x$ from $x - 7$ Answer ________
11. Subtract $7y - 3$ from $-7y - 3$ Answer ________
12. Subtract $-7x^2 + 3x - 7$ from $x^2 - 8x + 9$ Answer ________

13. Subtract $3x - 4$ from $x - 7$ Answer ________
14. Subtract $3x^2 - 7x + 6$ from $-3x^2 + 7x - 6$ Answer ________
15. Simplify $(7x^2 + 3x - 6) - (-x^2 + 6)$ Answer ________
16. Simplify $(-3x^2 + 6) + (3x - 9)$ Answer ________
17. Simplify $(2x^2 - 7x) + (-3x - 6)$ Answer ________
18. Simplify $(-3x^2 + 6x - 9) - (3x^2 + 6x - 9)$ Answer ________
19. $(4x + 8y) + (7x + 9y)$ Answer ________
20. Add $4a - 8b$ and $5a - 3b$ Answer ________
21. Add $7x^2 - 12x$ and $9x^2 - 11$ Answer ________
22. $(4a^2 - 7a) + (-3a + 9a^2)$ Answer ________
23. Add $-6x^2 + 9$ and $-3x + 11$ Answer ________
24. $(7b^2 - 6a) + (-3b^2 + 9a)$ Answer ________
25. Subtract $8y - 4$ from $-8y - 4$ Answer ________
26. Subtract $4x - 5$ from $x - 8$ Answer ________
27. Simplify $(-4x^2 + 7x + 10) - (-4x^2 + 7x - 10)$ Answer ________
28. Simplify $(-7x^2 - 7x - 10) - (7x^2 + 7x + 10)$ Answer ________

SECTION

6

Exponent Rule for Multiplication and Monomial Multiplication

Exponent Rule for Multiplication

1. When multiplying numbers with the same base, keep the base and add the exponents.

   $(A^m)(A^n) = A^{m+n}$

2. If no exponent is written, it is understood to be 1.

Monomial: Is a polynomial with only one term.
Binomial: Is a polynomial with two terms
Trinomial: Is a polynomial with three terms.
Polynomial: Is an algebraic expression in which all the exponents are whole numbers and in which there is no division by a variable (no variable in the denominator of a fraction).

Example 1: $c^7 \bullet c^2$

Solution: $c^7 \bullet c^2 = c^{7+2} + c^9$

Answer: c^9

Example 2: $a^7 \bullet a$

Solution: Notice $a = a^1$

$a^7 \bullet a = a^7 \bullet a^1 = a^{7+1} = a^8$

Answer: a^8

Fill in the correct answer in the space provided:

1. $(x^2)(x^3)$ Answer_________
2. $(x^2)(y^2)$ Answer_________
3. $(a^2)(a)$ Answer_________
4. $(a^7)(a^2)$ Answer_________

5. $(2^6)(2^5)$ Answer________

6. $(a^3)(a)$ Answer________

7. $(y^2)(y^7)$ Answer________

8. $(y^8)(x)$ Answer________

9. $(y^3)(x)$ Answer________

10. $(y^2)(y^7)$ Answer________

MULTIPLICATION OF A MONOMIAL BY A MONOMIAL

SECTION 7

RULES FOR MULTIPLICATION OF A MONOMIAL BY A MONOMIAL

1. Determine the sign of the product.
2. Multiply all the coefficients (the numbers in front of the unknowns).
3. Multiply the variable by adding exponents, if the bases are the same.

Example 1: $(-6x^2y^7)(3x^4y^2)$

Solution: Step 1: Sign of the answer $(-)$
Step 2: Multiply coefficients (numbers in front of the variables). (18)
Step 3: Multiplication: Same base – add exponents
$x^2 \cdot x^4 = x^6$
$y^7 \cdot y^2 = y^9$
$-18\ x^6\ y^9$

Answer: $-18\ x^6\ y^9$

Fill in the correct answer in the space provided:

1. $(-3x^2y)(-4xy^2)$ Answer________
2. $(7xy^2)(-2x^4y^6)$ Answer________
3. $(7rst)(-2rst)$ Answer________
4. $(-6x^2y^3)(2x^3y^8)$ Answer________
5. $(3ab^2c)(-4abc)$ Answer________
6. $(-4x^3)(-8x^2)$ Answer________
7. $(7xy^2z)(-2xyz)$ Answer________
8. $(3ab^2c)(-4a^2bc^2)$ Answer________
9. $(7rst)^2(-2r^2st)$ Answer________
10. $(-6x^2y^2z)(-2x^2y^2z^2)$ Answer________

SECTION

8

Multiplication of a Monomial by a Binomial

Distributive Law of Multiplication:

This law holds true for multiplication over addition. For any three numbers a, b, and c, multiplication before addition: $a(b+c) = ab + ac$.

Example 1: $3x^2(4x^3 - 6x)$

Solution: $3x^2(4x^3 - 6x)$ (Distributive Law)

$3x^2 \cdot 4x^3$

Answer: $12x^5 - 18x^3$

Example 2: $3a(a^2b + 4ab^2)$

Solution: $3a \cdot a^2b + 3a \cdot 4ab^2$

Answer: $3a^3b + 12a^2b^2$

For each of the products, fill in the correct answer in the space provided:

1. $3(x - 2)$ Answer________________
2. $-2(y - 4)$ Answer________________
3. $a(a + 4)$ Answer________________
4. $-b(b - 7)$ Answer________________
5. $-3x(x - 6)$ Answer________________
6. $4y(-y + 7)$ Answer________________
7. $-5z(2z^2 - 7z)$ Answer________________
8. $8x(-7x^2 - 8x)$ Answer________________
9. $5s^2(-2s + 7)$ Answer________________
10. $-4ab^2(-2a + 7)$ Answer________________
11. $7xy^2(3x^2y - 1)$ Answer________________
12. $-4ab^2(3a^2b + 3ab^2)$ Answer________________

13. $7a^2b(-4ab^2 + 3a^2b)$ Answer____________________

14. $-2xy^3(3x^3y^3 - 2x^2y^2)$ Answer____________________

15. $7x^2y(-3x^2 + 9y)$ Answer____________________

16. $-3xy(-4x^2y - 5y^2x)$ Answer____________________

17. $-4x(x - 7)$ Answer____________________

18. $3x^2(6x - 7)$ Answer____________________

SECTION

9

MULTIPLICATION OF A BINOMIAL BY A BINOMIAL

> Rules for Multiplication of a Binomial by a Bonomial:
> Hint: Multiply each term in the first binomial with each term in the second binomial.

Example 1: $(3x - 5)(2x - 1)$

Solution:

1. Multiply first term of each binomial to find the first term of the answer.
   $(3x)(2x) = 6x^2$
2. The middle term of the answer is found by adding the products of the two outside terms with the two inside terms in each binomial.
   $3x(-1) = -3x$ outside
   $-5(2x) = -10x$ inside
   $-13x$
3. The last term of the answer is found by multiplying the last term of each binomial.
   $(-5)(-1) = 5$

Answer: $6x^2 - 13x + 5$

This method is called the FOIL Method. The FOIL Method can only be used when you multiply a binomial by a binomial. What is FOIL?

F = First term

O = Outside term

I = Inside term

L = Last term

Let us look at Example 1 again:

$(3x - 5)(2x - 1)$

$F = (3x)(2x) = 6x^2$

$O = (3x)(-1) = -3x$

$I = (-5)(2x) = 5$

$L = (-5)(-1) = 5$

Let us put this together:

$6x^2 - 3x - 10x + 5$

Now combine like terms: $6x^2 - 13x + 5$ is the answer.

For each of the following products fill in the correct answer in the space provided.

1. $(x - 1)(x + 3)$
2. $(x - 4)(x + 5)$
3. $(2x - 3)(2x + 4)$
4. $(3x - 2)(2x + 1)$
5. $(y - 2)(y + 2)$
6. $(y - 5)(y + 5)$
7. $(x - 7)(x + 3)$
8. $(x - 5)(x + 6)$
9. $(3x - 6)(3x + 6)$
10. $(2x - 4)(3x + 6)$
11. $(x - 2)^2$
12. $(x + 5)^2$

SECTION

10

Multiplication of Polyonmials

Rules for Multiplication of Polynomials:

Hint: To find the product of two polynomials, use the column method just like in long multiplication with whole numbers.

Example: $(2x + 4)(3x^2 + 2x + 2)$

Solution:

$$\begin{array}{rrrrl} & 3x^2 & + 2x & + 2 & \\ & & 2x & + 4 & \\ \hline & 12x^2 & + 8x & + 8 & \leftarrow 4(3x^2 + 2x + 2) \\ 6x^3 & + 4x^2 & + 4x & & \leftarrow 2x(3x^2 + 2x + 2) \\ \hline \end{array}$$

Answer: $6x^3 + 16x^2 + 12x + 8$

For each of the products, fill in the correct answer in the space provided:

1. $(x + 2)(x^2 + 3x + 4)$
2. $(y - 4)(y^2 + 2y + 6)$
3. $(2x + 1)(x^2 - 2x + 4)$
4. $(y + 3)(y^2 - 3y + 9)$
5. $(x - 2)(x^2 + 4x + 6)$
6. $(y - 3)(y^2 - y + 4)$
7. $(5y^2 + 2y + 2)(y^2 - 3y + 5)$
8. $(y - 3)(y^2 - y + 4)$

9. $(5y^2 + 2y + 2)(y^2 - 3y + 5)$
10. $(2x^2 + x + 1)(x^2 - 4x + 3)$
11. $(Y - 5)(Y^2 + 3Y + 7)$
12. $(3X - 6)(X^2 - 4X + 7)$

SECTION 11

Evaluating Algebraic Expressions

Rules for Evaluating Algebraic Expressions

1. Substitute the value for the unknown in the given expression.
2. Evaluate the roots and powers.
3. Multiply or divide from left to right.
4. If necessary, add or subtract.

Example 1: find the value of $3a - 1$ when $a = 6$.

Solution: $3a-1$

$3(6)-1$ Substitute $a = 6$

$18-1$ Order of operations

Answer: 17

Example 2: Find the value of $a^2 + 7a$, when $a = -2$

Solution: $a^2 + 7a$

$(-2)^2 + 7(-2)$ Substitute $a = -2$

$4 + (-14)$ Order of operations

-10

Answer: -10

Example 3: $y^2 + 3x^2$, when $x = -1$ and $y = 2$

Solution: $y^2 + 3x^2$

$(2)^2 + 3(-1)^2$ Substitute $x = -1$ and $y = 2$

$4 + 3(1)$ Order of operations

$4 + 3$

7

Answer: 7

Example 4: Find the value of $x^3y + 4x$, when $x = -2$ and $y = 5$

Solution: $x^3 + 4x$

$(-2)^3(5) + 4(-2)$ Substitute $x = -1$ and $y = 5$

$(-8)(5) + 4(-2)$ Order of operations

$-40-8$

-48

Answer: -48

Example 5: If $a = -3bc$, find a, when $b = -2$ and $c = -3$

Solution: $a = -3bc$

$a = -3(-2)(-3)$

$a = 6(-3)$

$a = -18$

Answer: -18

Fill in the correct answer in the space in the provided:

1. Find the value of $3x-4$, when $x = -2$
2. Find the value of $a^2 - 7a$, when $a = 4$
3. Evaluate $-b^2$ when $b = -5$
4. Evaluate $t^2 - 7t + 6$ when $t = 8$
5. Find the value of $3a + 5b$ when $a = -6$ and $b = -7$
6. Find the value of $4c - 5d$ when $c = 2$ and $d = 3$
7. Find the value of $6x^2 + 7y^2$ when $x = 3$ and $y = 4$
8. Find the value of $x^2 + 2xy$ when $x = 3$ and $y = 4$
9. If $a = 3bc$, find "a", when $b = 4$ and $c = -5$
10. If $x = 3y^2$, find "x", when $y = 2$ and $y = 3$.
11. Evaluate $-3x^2$, when $x = 4$
12. Evaluate $7x^2 - 4y^2$, when $x = 2$ and $y = 3$.
13. Find the value of $x^2 + 6y$, when $x = 4$ and $y = 5$
14. Evaluate $-4x^2$ when $x = -5$

Order of Operations in Algebra

SECTION 12

To Simplify:
Hint: First perform multiplication and then addition and subtraction

Example 1: Simplify $3x + x(x + 6)$

Solution: $3x + x(x + 6)$
$3x + x^2 + 6x$
$x^2 + 9x$
Answer: $x^2 + 9x$

Distribution Law of Multiplication
Addition of like terms

Example 2: Simplify $3ab^2 + 4ab(6b-7)$

Solution: $3ab^2 + 4ab(6b - 7)$
$3ab^2 + 24ab^2 - 28ab$
$27ab^2 - 28ab$
Answer: $27ab^2 - 28ab$

Distribution Law of Multiplication
Addition of like terms

Example 3: Simplify $-3x^2 + 2x(x - 7)$

Solution: $-3x^2 + 2x(x - 7)$
$-3x^2 + 2x^2 - 14x$
$-x^2 - 14x$
Answer: $-x^2 - 14x$

Simplify each of the following example and write the correct answer in the space provided:

1. $3y + y(y + 7)$
2. $2x + x(x + 4)$
3. $3x + x^2(x - 7)$
4. $4y + 4y^2(-y + 2)$
5. $6x^2 - 7x(x - 4)$
6. $-3y^2 + 4y^2(y + 2)$

7. $7x^2 + 2x^2(-x - 4)$
8. $3y^2 + 4y^2(-y - 7)$
9. $3a^2b - 2a(ab - 7)$
10. $x^2y - 7x(x^2 + 3xy)$
11. $3x^2y^2 - 7x(xy - 7xy^2)$
12. $-6xy - 3xy(-2x - 5y)$
13. $6a^2 - 3ab(7a - 2b)$
14. $4x^2y - 3xy(-2x - 5y)$
15. $5y + y(y + 9)$
16. $5x + x(x - 9)$
17. $7x^2 + 4x^2(-x - 7)$
18. $3y^2 + 5y^2(y + 5)$
19. $3x^2 - 2x(x + 4)$
20. $4y^2 + 6y^2(y + 6)$

Problem Set #3

Interpret each of the following algebraic expressions:

1. Six more than a number
2. A number minus eight
3. A number decreased by six
4. A number plus seven
5. 8 times a number
6. 5 less than a number
7. A number increased by 14
8. 4 more than one third a number
9. 5 more than 6 times a number
10. 7 less than 4 times a number
11. Find the value in cents of "q" quarters and "d" dimes.
12. Find the value in cents of "n" nickels and "q" quarters.

Problem Set #4

Add the polynomials:

1. $(3x + 12) + (-3x - 2)$
2. $(5x - 2y) + (3x + 8y)$

3. $(8x^2 + 5y^2 - 10z) + (3x^2 + 2y^2 + 8x)$
4. $(y^2 - y - 5) + (3y - 2y^2 + 1)$
5. $(4x^3 + 5x - 7 - 4x^2) + (4x^2 - 2x^3 + 4x - 1)$
6. $(4a^2 + 6a - 7) + (4a - 2a^2b + 5ab^2)$

Problem Set #5

Subtract the polynomials

1. $(2x - 3y + 2) - (x + y - 7)$
2. $(x^2 - 4x) - (x - 2)$
3. $(5x^2 + 2) - (3 - x^2)$
4. $3xy - (4x - 2xy)$
5. $(5x^3 - 6x^2 + 2x - 7) - (2x^3 - 6x - 1)$
6. $(-10xy^2 + 5x^2y) - (4xy^2 - 3x^2y)$

Problem Set #6

Multiply the following:

1. $6(a-b)$
2. $2(x^2 - y - 3z)$
3. $xy(x - 2y)$
4. $7x^2(x^2 - 3x - 5)$
5. $-xy^2(2x^2y - 3x^2)$
6. $-5xyz(-2xy - 3xz - 4y^2)$
7. $(x - 3)(2x + 4)$
8. $(6x - 3)(3 - x)$
9. $(4x - 1)(3x + 5)$
10. $(x - 3)^2$
11. $(3x - y)^2$
12. $(x + 2y)(x - 3y)$
13. $(x - y)^2$
14. $(x + 3)(x^2 + 4x + 5)$
15. $(x - 5)(x^2 + 3x - 7)$
16. $(x - 4)(x^2 - x - 4)$

Problem Set #7

Evaluate each algebraic expression:

1. Find the value of $3x - 5$ when $x = -3$
2. Find the value of $A^2 - 8A$, when $A = -5$
3. Evaluate $-b^2$ when $b = -6$
4. Find the value of $4A + 5B$ when $A = -6$ and $B = -7$
5. Find the value $x^2 + 7xy +$ when $x = 4$ and $y = 5$
6. Find the value of $4x^3 + 7x^2$ when $x = -1$
7. Evaluate $-3x^2$ when $x = -4$
8. Evaluate $-3x^3y^2$ when $x = -2$ and $y = -3$
9. Find the value of $x^2 + 6y$ when $x = 5$ and $y = 6$
10. Evaluate $-5x^2$ when $x = -6$

Problem Set #8

Evaluate each algebraic expression after setting $x = 2$, $y = -3$, $z = -4$

1. $x - y - (x - z)$
2. $x^2 - 3xy - z$
3. $2xy - 3(x - z)$
4. $xz^2 - (y - z)$
5. $x^2 - 4xy$
6. $x - (y - z)^2$

Problem Set #9

Using the order of operations in algebra simplify each expression:

1. $4y + y(y + 8)$
2. $3x + x(x - 5)$
3. $4x + x^2 (x - 7)$
4. $5y + 5y^2(-y + 10)$
5. $7x^2 - 8x(x - 5)$
6. $4x^2 + 5x^2(-x - 8)$
7. $2x + x(x - 9)$
8. $3x^2 + 6x(x - 7)$

SECTION

13

More on the Rules of Exponents — Exponent Rule

Exponent Rule: Raising a number to a Power

To raise a number to a power, multiply the exponents

$(x^m)^n = x^{mn}$

Example 1:

Solution: $(a^4)^2 = a^4 \cdot a^4 = a^8$

Answer: a^8

Example 2: $(x^2y^4) = x^6y^{12}$

Solution: $(x^2y^4)^3 = x^{2*}3y^{4*}3 = 125x^6y^{13}$

Answer: $125x^6y^{12}$

Fill in the letter of the correct letter in the space provided:

1. $(x^4)^3$
   (a) x (b) x^7 (c) x^{12} (d) x^5 (e) x^{64}
2. $(y^{12})^2$
   (a) y^{10} (b) y^{24} (c) y^{14} (d) y^{144} (e) y^{36}
3. $(x^7)^2$
   (a) x^{14} (b) x^{49} (c) x^9 (d) x^{144} (e) x^{17}
4. $(y^{15})^4$
   (a) y^{19} (b) y^{11} (c) y^{10} (d) x^5 (e) y^{36}
5. $(ab^2)^3$
   (a) a^2b (b) $a^3\ b^8$ (c) ab^6 (d) a^2b^2 (e) a^3b^6
6. $(c^2d^4)^5$
   (a) $c^{10}d^{20}$ (b) $c^{15}\ d^7$ (c) c^3d (d) c^3d^9 (e) c^4d^7
7. $(c^5d^7)^3$
   (a) $16b^4$ (b) $c^{15}d^7$ (c) c^5d^4 (d) c^2d^{21} (e) $c^{15}d^{21}$
8. $(2x^2\ y^3)^4$
   (a) $16x^8y^{12}$ (b) $16x^6y^7$ (c) $64\ x^8y^{12}$ (d) $64\ x^2y^{12}$ (e) 16^8y^7

SECTION 14

Division Of Algebraic Expressions

Exponent Rule For Division

When dividing numbers with the same base, keep the base and subtract the exponents. (exponent in the numerator minus exponent in the denominator)

Example 1: $\dfrac{C^7}{C^2}$

Solution: $C^{7-2} = C^5$

Example 2: $\dfrac{a^7}{a}$

Solution: $\dfrac{a^7}{a^1} = a^{7-1} = a^6$

Rules: for Division in Algebra by a Monomial

1. Divide each part of the numerator by the denominator.
2. Determine sign of quotient.
3. Divide the coefficients.
4. Subtract exponents – same base.

Example 1: $\dfrac{8a^2}{4a}$

Solution: Divide coefficients and subtract exponents

Answer: $2a$

Example 2: $\dfrac{6b^4 - 12}{3} = \dfrac{6b^4}{3} - \dfrac{12}{3} = 2b^4 - 4$

Solution: $2b^4 - 4$

Fill in the correct answer in space provided:

1. $\frac{6x^2}{2x}$ Answer:________

2. $\frac{6a - 12}{3}$ Answer:________

3. $\frac{10y - 20}{-5}$ Answer:________

4. $\frac{-4x^2}{-x^2}$ Answer:________

5. $\frac{6a^2 - 3a}{-3a}$ Answer:________

6. $\frac{8y^3 - 6y^2}{-4y}$ Answer:________

UNIT *two*

Section 15: Solving Linear Equations 42

Section 16: Solving Equations with Fractions 44

Section 17: Solving Literal Equations 49

Section 18: Solving Verbal Problems—Number Problems 52

Section 19: Ratios and Proportions—Verbal Problems 55

Section 20: Inequalities 59

Section 21: Absolute Value 62

Section 22: Relations and Functions 65

Section 23: Graph of a Straight Line 68

Section 24: Solving Simultaneous Equations – Algebric Solutions Graphic Solutions 77

Section 25: Properties of Exponents 86

Section 26: Factoring A Polynomial 89

Section 27: Factors: The difference of Two Squares 91

Section 28: Trinomial Factors 93

Section 29: An Additional Factoring Problem 97

Section 30: Rational Expressions 101

Section 31: Complex Numbers 106

Section 32: Solving Quadratic Equations 112

Section 33: To Solve Quadratic Equations by using the Quadratic Formula 114

Section 34: Radical Equations 119

SECTION 15

Solving Linear Equations

Rules: Remember we saw this earlier

1. An equation is statement of equality. There are three parts to an equation: the right side, the left side, and the equal sign.
2. To solve the equation, use the addition property, Addition Inverse: for every number A, there is exactly one number (−A), such that:

   $A + (-A) = 0$, $X - 2 = 3$
   $X = 5$ (add 2 to both sides)

3. If necessary, in the last step, use the Multiplication Property.
   Multiplication inverse; It is the product of two numerals or two variables that = 1.

   $N \bullet 1/N = 1$
   $2X = 6$
   $X = 3$

   Multiply both sides by the multiplicative inverse of 2, which is ½.

Example 1: $3x - 6 = 12$

Solution: $3x - 6 = 12$

$3x - 6 + 6 = 12 + 6$ Additive Inverse

$1/3 \bullet 3x = 18 \bullet 1/3$ Multiplicative Inverse

Answer: $x = 6$

Example 2: $3(x + 7) = 21$

Solution: $3(x + 7) = 21$

$3x + 21 = 21$

$3x + 21 - 21 = 21 + 21$

$1/3 \bullet 3x = 42 \bullet 1/3$

Answer: $x = 14$

Example 3: $3x + 6 + 2x = 31$

Solution: $3x + 6 + 2x = 31$

$5x + 6 = 31$

$5x + 6 - 6 = 31 - 6$

$1/5 \bullet 5x = 25 \bullet 1/5$

Answer: $x = 5$

Example 4: $6x + 1 = 2x + 45$

Solution: $6x + 1 = 2x + 45$

$6x + 1 - 1 = 2x + 45 - 1$

$6x = 2x + 44$

$6x - 2x = 2x - 2x + 44$

$1/4 \bullet 4x = 44 \bullet 1/4$

Answer: $x = 11$

Solve each of the following equations:
Write the correct answer in the space provided.

1. $x - 5 = 2$ Answer: ________
2. $x + 7 = 4$ Answer: ________
3. $2x = 6$ Answer: ________
4. $-3y = 9$ Answer: ________
5. $2y - 7 = 9$ Answer: ________
6. $-3x - 6 = 9$ Answer: ________
7. $-3x + 6 = 12$ Answer: ________
8. $4 - x = 0$ Answer: ________
9. $3x + 4 = 2y + 6$ Answer: ________
10. $y - 1 = 7$ Answer: ________
11. $2x - 6 = -12$ Answer: ________

SECTION 16

SOLVING EQUATIONS WITH FRACTIONS

Rules:

1. Multiply each and every number and variable on both sides of the equation by the lowest common denominator(LCD).
2. Now follow the rules used for solving equations.

Example 1: $\frac{a}{6} + 12 = \frac{a}{3}$

Solution: The LCD for $\frac{a}{6}$ *and* $\frac{a}{3}$ is 6.

Multiply both sides by 6.

$\frac{a}{6} + 12 = \frac{a}{3}$

$6\left(\frac{a}{6}\right) + 6(12) = 6\left(\frac{a}{3}\right)$

$a + 72 = 2a$

$a - a + 72 = 2a - a$

$72 = a$

Answer: $a = 72$

Example 2: $\frac{x}{3} + \frac{x}{4} = 12$

Solution: The LCD for $\frac{x}{3}$ *and* $\frac{x}{4}$ is 12

Multiply both sides by 12.

$\frac{x}{3} + \frac{x}{4} = 12$

$12\left(\frac{x}{3} + \frac{x}{4}\right) = 12 \times 12$

$4x + 3x = 144$

$$x = \frac{144}{7}$$

Answer: $x = \dfrac{144}{7}$

Solve each of the equations. Write the answer in the space provided.

1. $\dfrac{x+1}{2} = \dfrac{3x+6}{4}$

2. $\dfrac{x-4}{3} = \dfrac{2x+1}{5}$

3. $\dfrac{x}{2} + \dfrac{x}{4} = 3$

4. $\dfrac{x}{3} + \dfrac{x}{9} = 4$

5. $\dfrac{x}{12} + 1 = \dfrac{x}{10}$

6. $\dfrac{x}{2} + 3 = \dfrac{x}{4}$

7. $\dfrac{u}{10} = \dfrac{u}{5} - 2$

8. $\dfrac{x}{7} + 2 = \dfrac{x}{14}$

9. $\dfrac{x}{2} + \dfrac{x}{3} = 5$

HINTS TO SOLVING AN EQUATION

Rules:

1. Remove all denominators by multiplying each item by the LCM (Least Common Multiple).
2. Remove all parentheses by using the distributive law.
3. Simplify both sides of the equation.
4. Bring the letter to one of side of the equation and bring the number to the opposite side of equation and simplify both sides.(This is the inverse law for addition.)
5. Divide both sides of the equation by whatever is in front of the letter. (This is the inverse for multiplication.) You just solve for the unknown letter.

Problem Set 12

1. $x - 6 = 4$
2. $x + 8 = 6$
3. $-3x = 12$
4. $-4y = -16$
5. $2y - 9 = 15$
6. $-3x - 6 = -12$
7. $5 - y = 0$
8. $6 - 7y = 20$
9. $3x - 6 = 2x + 15$
10. $4x - 7 = x + 8$
11. $3x + 6 + 2x = 31$
12. $4x - 7 - x = 2$
13. $2y - 9 = 5$
14. $-3(x - 6) = 18$
15. $-2(x + 4) = -8$
16. $7x - 9 = x + 15$
17. $7x - 2 = 2x + 23$
18. $3(x - 7) = -3$
19. $3x + 4 = 2x + 6$
20. $2y - 6 = -12$
21. $3x - 7 = 2x + 5$
22. $3x - 5 = 7$
23. $3(x - 5) = 7$
24. $70 = 6x + 10$
25. $-5y + 17 = 47$
26. $x + 3 = 4x + 1$
27. $2x - 5 = x - 8$
28. $2(x + 5) = 10 - x$
29. $5(y - 2) = -2(y - 2)$
30. $6x + 3x - 12 = 2x$
31. $9x + 3(2x - 1) = 12$
32. $3(y - 2) - 2 = 5(y + 3) - 7(y - 1)$

33. $2(x-1)+2+3\left(x-\frac{1}{3}\right)+1=-4\left(x-\frac{9}{4}\right)$

34. $-2x-3(2x-4)=4x+8$

35. $\frac{1}{3}x-3=4$

36. $y-\frac{1}{4}=5\frac{3}{4}$

37. $2(x+1)=7x-8$

38. $z-\frac{5}{4}=\frac{12}{2}$

39. $\frac{3x}{4}-6=\frac{x}{12}$

40. $\frac{x}{3}+\frac{x}{4}=22$

41. $\frac{x}{3}-\frac{3x}{8}=5+\frac{3x}{4}$

42. $\frac{3x-2}{5}=1$

43. $\frac{1-2x}{4}=2$

44. $\frac{y}{9}+\frac{y}{3}=4$

45. $\frac{y}{2}+3=\frac{y}{4}$

46. $\frac{x}{3}+2=6$

47. $\frac{x}{5}=\frac{x}{10}-4$

48. $\frac{y}{3}+\frac{y}{2}=5$

49. $\frac{y}{24}+2=\frac{y}{20}$

50. $\frac{x}{3}+1=\frac{x}{2}+6$

51. $\frac{y+1}{4}=\frac{3y+6}{2}$

52. $\frac{x+7}{4}=\frac{3x-6}{2}$

53. $\frac{x}{3} + \frac{x}{4} = 12$

54. $\frac{x-7}{3} = \frac{2x+7}{4}$

55. $\frac{x}{4} - 7 = 14$

SECTION

17

SOLVING LITERAL EQUATIONS

Rules:

1. A literal equation is an equation with two or more variables.
2. Isolate the variable to be solved.
3. Follow the rules used to solve linear equations.

Example 1: If $3y - 7 = x$, solve for y.

Solution: To solve for y, the variable y must be isolated.

$3y - 7 = x$

$3y - 7 + 7 = x + 7$

$\frac{1}{3} \times 3y = \frac{1}{3}(x + 7)$

$y = \frac{x + 7}{3}$

Answer: $y = \frac{x + 7}{3}$

Example 2: If $bx + a = 3c$, solve for x.

Solution: Isolate x on the left side. Use Additive Inverse by adding $-a$ to both sides of the equation.

$bx + a = 3c$

$bx + a - a = 3c - a$ Additive Inverse

$\frac{1}{b} \bullet bx = \frac{1}{b} \bullet (3c - a)$ Multiplicative Inverse

Answer: $x = \frac{3c - a}{b}$

Solve each of the following equation for the variable indicated. Write the correct answer in the space provided.

1. $x - y = 7$ Solve for x: Answer: ________
2. $x - y = 3$ Solve for y: Answer: ________
3. $a - 7b = 0$ Solve for a: Answer: ________
4. $a - 2b = 0$ Solve for b: Answer: ________
5. $3x - 2 = y$ Solve for x: Answer: ________
6. $4x - 5y = 6$ Solve for y: Answer: ________
7. $\frac{x}{2} + 4 = y$ Solve for y: Answer: ________
8. $\frac{x}{3} - 7 = y$ Solve for x: Answer: ________
9. $x + y + z = 6$ Solve for z: Answer: ________
10. $3x - 4y + 7z = 9$ Solve for y: Answer: ________
11. $\frac{a - b}{4} = c$ Solve for a: Answer: ________
12. $3(x + y) = 6$ Solve for x: Answer: ________
13. $-2(x - y + z) = 12$ Solve for x: Answer: ________

Problem Set 13

Solving Literal Equations.

1. $x - y = 9$ Solve for x: Answer: ________
2. $a + 8b = 0$ Solve for a: Answer: ________
3. $4x - 5 = y$ Solve for x: Answer: ________
4. $5x - 6y = 7$ Solve for y: Answer: ________
5. $\frac{x}{3} + 4 = y$ Solve for x: Answer: ________
6. $\frac{x}{4} - 8 = y$ Solve for x: Answer: ________
7. $x - 2y = 09$ Solve for y: Answer: ________
8. $A + B + C = 0$ Solve for B: Answer: ________
9. $4x - 5y + 8z = 3$ Solve for x: Answer: ________
10. $4(x + y) = 6$ Solve for y: Answer: ________
11. $\frac{x - y}{3} = c$ Solve for x: Answer: ________

12. $\frac{A-B}{2}=d$ Solve for a: Answer: ________

13. $\frac{x}{4}-8=y$ Solve for y: Answer: ________

14. $2a-3b=4c$ Solve for a: Answer: ________

15. $F=MA$ Solve for M: Answer: ________

16. $r^2=x^2+y^2$ Solve for y^2: Answer: ________

17. $z=\frac{x-m}{s}$ Solve for x: Answer: ________

18. $y=mx+b$ Solve for b: Answer: ________

19. $A=P(1+r)$ Solve for P: Answer: ________

20. $V=mr^2h$ Solve for h: Answer: ________

21. $a(c-d)=b$ Solve for a: Answer: ________

22. $P=2(L+W)$ Solve for L: Answer: ________

23. $A=\frac{1}{2}(bh)$ Solve for h: Answer: ________

SECTION 18

SOLVING VERBAL PROBLEMS—NUMBER PROBLEMS

Rules:

English	Algebra
The sum of x and y	$x + y$
The product of x and y	xy
The quotient of x and y	$\frac{x}{y}$
A number	x
6 more than a number	$6 + x$
4 times a number	$4x$
The difference of x and y	$x - y$

Example: The sum of twice a number and four is eight. Find the number.

Solution:
1. Let $x =$ the number
2. The sum of twice a number and four is represented as: $2x + 4$.
3. The word "is" translated as $=$.
4. Eight is 8.
5. Equation: $2x + 4 = 8$
6. Solve for x.

$2x + 4 = 8$ Add -4 to both sides Additive Inverse

$2x = 4$ Multiply both sides by ½

$x = 2$

Answer: $x = 2$

Some other key words are:

Is mean equal to

Of means to multiply

What is the variable you are solving for

Let us look at an example using these words:

Example 1: What is 20% of 80?
X = 20% of 80
X = 0.20(80)
X = 16

Example 2: Forty is what percent of eighty?
40 = x(80)
Let us clean this up. You then get:

$$\frac{80x}{80} = \frac{40}{80}$$

x = ½
x = 50%

Example 3: Fifty percent of what number is 200?
50%(x) = 200
50%x = 200
0.50x/0.5 = 200/0.5
x = 400

Solve the following problems. Write the correct answer in the space provided.

1. The sum of twice a number and six is twelve. Find the number.
   Answer:__________
2. Four times the sum of a number and seven is thirty. Find the number.
   Answer:__________
3. One number is four less than another. Their sum is eight. Find both numbers.
   Answer:__________
4. The sum of twice a number and eight is four. Find the number.
   Answer.__________
5. If twice the difference of a number and three were decreased by five, the result would be three. Find the number.
   Answer:__________
6. One number is 6 less than another. Their sum is twelve. Find the larger number.
   Answer:__________
7. Five times the sum of a number and six is forty. Find the number.
8. The sum of twice a number and nine is fifteen. Find the number.
   Answer:__________

Problem Set 14

Solve the following problems. Set up an equation for each problem. Only algebraic solution accepted. Show all work.

1. The sum of four times a number and seven is sixty-seven. Find the number.
2. One number is six less than another. Their sum is twelve. Find the number.
3. The sum of twice a number and sixteen is twenty-four. Find the number.
4. Six times a number and seven is forty-three. Find the number.
5. One number is eight less than another. The sum is fifty-eight. Find the smaller number.
6. A number divided by ten is thirty. Find the number.
7. Two more five times a number is thirty-seven. Find the number.
8. One number is ten less than another; their sum is fifty. Find the number.
9. At a baseball game, each ticket costs $10. If expenses total $3,000, how many tickets must be sold in order to have a profit of $2000?
10. It costs a manufacturer $20 to produce each video game. In addition, there is a general overhead of 30,000. He produces 6,000 games. If he received $50 per game from a wholesaler, how many games must he sell to break even?
11. A number increased by 3 is 15. Find the number.
12. A number decreased by 5 is 25. Find the number.
13. A number increased by 10 is 32. What is the number?
14. A number diminished by 4 is 28. Find the number.
15. 12 more than a number is 32. What is the number?
16. When 15 is added to a number, the result is 27. What is the number?
17. When 15 is subtracted from a number, the result was 27. What is the number?
18. The total cost of a shirt is $16. If the shirt costs three times as much as the neck tie, what is the cost of each?
19. The sum of three consecutive integers is 69. Find the number.
20. A man distributed $310 among his three friends, Tom, Sammy, and Mike. He gave Sammy 3 times as much as Tom, and Mike $10 more than Tom. How much money did each receive?
21. The greater of two numbers is two more than three times the smaller. Four times the greater exceeds five times the smaller by 22. Find the numbers.
22. Mrs. Jones bought two kinds of candy for a party; one kind costing 40 cents a pound, and the other 70 cents a pound. If she pays $3.40 for 7 pounds of candy, how many pounds of each kind did she buy?
23. Perry is twice as old as Jim. Three years ago Perry was three times as old as Jim was then. What are the present ages of both boys?
24. John sold 500 tickets to a football game and collected a total of $600. If he charged students 75 cents per ticket and non students $1.50 per ticket, how many students did he sell ticket to?

SECTION 19

Ratios and Proportions—Verbal Problems

1. Ratio: If x and y are any two numbers, where $y \neq 0$, then the ratio of x and y is: $\frac{x}{y}$
2. Proportion: is a statement that two ratios are equal. Means—extremes property. If a, b, c, d are real numbers, when $b \neq 0$ and $d \neq 0$, then if $\frac{a}{b} = \frac{c}{d}$, then $ad = bc$.
3. In words: In any proportion, the product of the means is equal to the product of the two extremes.

Example 1: Solve $\frac{3}{x} = \frac{6}{7}$

Solution: 1. $\frac{3}{x} = \frac{6}{7}$ Extremes are 3 and 6
Means are x and 7

2. $6x = 21$

3. $x = \frac{21}{6}$ Multiply both sides by 1/6

4. $x = \frac{7}{2}$

Answer: $x = \frac{7}{2}$

Example 2: A baseball player gets 6 hits in the first 18 games. If he continues hitting at the same rate, how many hits will he get in the first 45 games?

Solution: 1. Ratio 1 = Ratio 2

2. $\frac{Hits}{Games} = \frac{Hits}{Games}$

3. $\frac{6}{18} = \frac{x}{45}$

4. 270 = 18x — Product of extremes = product of means

5. 15 = x — Multiply both sides by 1/18, Multiplicative Inverse

Answer: x = 15

Solve each of the following proportions.
Write the correct answer in the space provided:

1. $\frac{x}{3} = \frac{6}{9}$ Answer:____________

2. $\frac{4}{x} = \frac{2}{3}$ Answer:____________

3. $\frac{15}{60} = \frac{60}{x}$ Answer:____________

4. $\frac{2}{7} = \frac{x}{14}$ Answer:____________

5. $\frac{x+2}{3} = \frac{2}{5}$ Answer:____________

6. $\frac{x+1}{4} = \frac{2}{9}$ Answer:____________

Problem Set #15

1. If 200 grams of ice creams contain 26 grams of fat, how much fat is in 700 grams of ice cream?
2. A map is drawn so that 2 inches represents 350 miles. If the distance between 2 cities is 875 miles, how far apart are they on the map?
3. An airplane flies 1950 miles in 6 hours. Har far will it travel in 7 hours?
4. A man drives his car 630 miles in 10 hours. At this rate, how far will he travel in 12 hours?
5. A 5-ounce serving of grapefruit juice contains 125 grams of water. How many grams of water are there in 8-ounce serving of grapefruit juice?
6. A basketball player makes 8 out of 12 free throws in the first game of the season. If she shoots with the same accuracy in the second game, how many of the 15 free throws she attempts will she make?
7. If 400 grams of candy contains 68 grams of fat, how many grams of fat are there in 700 grams of candy?
8. Zelma drives her car 720 miles in 8 hours. At this rate, how far will she travel in 11 hours?
9. A map is drawn so that 2 inches represents 700 miles. If the distance between 2 cities is 3,850 miles, how far apart are they on the map?

10. A 7-ounce serving of orange juice contains 105 grams of water. How many grams of water are there in 12-ounce serving of orange juice?

11. A baseball player makes 2 hits every 9 times at bat. If he hits the ball with the same accuracy, how many hits would he get if he were at bat 450 times?

PROBLEM SET #16

1. $4x + 7 = 2x + 9$
2. $3(8x + 4) = 2(x - 3)$
3. $\frac{3}{4}x + \frac{7}{8} = -1$
4. $\frac{7}{8}x + \frac{1}{4} = 8x + 9$
5. $\frac{2}{5}x = \frac{8}{10}$
6. $3(x - 4) = 12$
7. $7x + 4 = 2x + 29$
8. $3x + 7 - x = 12 + 11$
9. $-7(x + 4) = 56$
10. $3x - 7 - 7x = 21 + 7$
11. $\frac{3}{4}x + \frac{1}{2} = \frac{2}{3}$
12. $2(x - 3) - 5 = 4(x - 5)$
13. $3x - 10 = 11$
14. $-3y - 12 = -24$
15. $\frac{12 + x}{-4} = \frac{-7 + 5x}{3}$
16. $\frac{1}{2}x + 7 - \frac{1}{4x} = \frac{19}{2}$
17. $-3x - 7 = -2(x + 4)$
18. $-\frac{3}{4}x = \frac{-6}{8}$
19. $3(x + 4) = -48$
20. $-3(x + 4) = -2(x + 7)$
21. The sum of three times a number and six is eighteen. Find the number.
22. One number is six less than another. Their sum is eighteen. Find both numbers.
23. Five times the sum of a number and eight is one hundred and twenty. Find the number.

24. The sum of twice a number and eighteen is thirty. Find the number.

25. A man travels 1260 miles in 20 hours. At this rate, how far will he travel in 12 hours?

26. A jet airplane flies 3,900 miles in 12 hours. How far will it travel in 14 hours?

27. If 400 grams of ice cream contains 52 grams of fat, how much fat is in 1,400 grams of ice cream?

28. A map is drawn so that 4 inches represents 700 miles. If the distance between 2 cities is 1,750 miles, how far apart are they on the map?

29. A 15 ounce serving of tomato juice contains 375 grams of water. How many grams of water are there in a 24 ounce serving of tomato juice?

30. If 50 grams of chocolate pudding contains 12 grams of fat. How many grams of fat are there in 175 grams of chocolate pudding?

SECTION

20

INEQUALITIES

A mathematical statement that one quantity of an expression is less than or greater than another is an inequality.

Example 1: $7 > 4$

The above expression states that 7 is greater than 4.

Example 2: $3 < 5$

The above statement states that 3 is less than 5.

Symbols		
1.	$>$	Greater than
2.	$<$	Less than
3.	$\geq$	Greater than or equal to
4.	$\leq$	Less than or equal to

SOLVING INEQUALITIES:

An inequality is a true statement provided a **solution set** of all real numbers is found. For example: the solution set of $x + 3 > 6$ is the set of all real numbers greater than 3. If two inequalities have the same solution set, they are called equivalent inequalities. To solve inequalities, simpler and equivalent inequalities must be found.

The following properties of inequalities must be applied in order to find a solution.

PROPERTIES OF INEQUALITIES

Let X, Y, Z represent real numbers.

1. Addition—Subtraction Property:
   If the same real number is added to or subtracted from both sides of the inequalities the

inequality symbol separates the two sides, the resulting inequality is equivalent to the original inequality.

$X \leq Y$ and $X + Z < Y + Z$ are equivalent inequalities

2. Multiplication—Division Property:

   a. An equivalent inequality is found when one multiplies or divides each side by the same positive real number.

      If $Z > 0$, then $X < Y$ and $XZ < YZ$ are equivalent inequalities.

   b. An equivalent inequality is found when one multiplies or divides each side by the same negative real number provided the direction of the inequalities symbol is reversed.

      If $Z < 0$, then $X < Y$ and $XZ > YZ$ are equivalent inequalities.

In essence, an equivalent inequality is produced when multiplying or dividing by the same positive or negative real number. However, the direction of the inequality must be reversed when multiplying or dividing by a negative number.

Example 1: $3x + 2 < 8$

$3x$	< 6	Add -2 to both sides
x	< 2	Divide both sides by 3
		Keep the same inequality symbol

Example 2: $-4x - 8 \leq 12$

$-4x$	≤ 20	Add 8 to both sides
x	≥ -5	Divide both sides by -4 reverse
		the direction of the inequality symbol.

Use the properties of inequalities to solve each of the following. Write your answer in the space provided.

1. $2x + 4 > 12$
2. $3x - 7 < 8$
3. $-3x + 7 \leq 10$
4. $-4x - 8 \leq -16$
5. $x + 5 > 3x + 17$
6. $3x - 7 \leq x + 9$
7. $-3(x + 2) > 5x + 7$
8. $3x - 5 > 16$
9. $-7(x + 4) \leq 35$
10. $3x - 7 + 2x < 8$
11. $-4(3x - 5) > 2(x + 4)$
12. $-3(x + 7) < -4(x + 9)$
13. $-2(x - 9) \geq 18$

14. $3(x - 7) \leq 21$

15. $x + 4 \geq 5x + 16$

16. $-6(x + 2) = x + 4$

17. $-x + 6 \geq -12 + 3x$

18. $3x - 3 + x > 8$

19. $-4(x + 1) \leq 12$

20. $-7(x + 4) < -3(x + 1)$

Absolute Value

SECTION 21

The **Absolute Value** of the real number X, represented by |X| is the distance between X and 0 on a number line.

For example: $|4| = 4$ and $|-4| = 4$ because both numbers 4 and -4 are 4 units from zero.

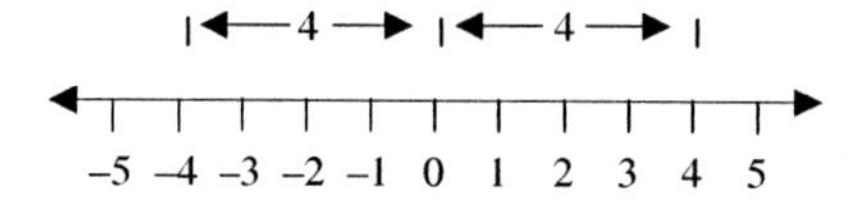

In general, $X \geq 0$ then $|X| = X$. If $X < 0$, then $|X| = -X$ because $-X$ is positive when $X < 0$.

The following definition is derived:

Definition of Absolute Value:

The absolute value of the real number X is defined by:

$$|X| = \begin{cases} X, & \text{if } X \geq 0 \\ -X, & \text{if } X < 0 \end{cases}$$

The distance of two points on a real number line makes use of absolute value.

The following property is used to solve Absolute Value Equations:

Property of Absolute Value Equations:

For any variable expression E and any non-negative real number K,
$|E| = K$ if and only if $E = K$ or $E = -K$

Example 1:

Solve: $|2x - 6| = 22$

Solution:

$2x - 6 = 22$	or	$2x - 6 = -22$	(Definition of Absolute Value)
$2x = 28$		$2x = -16$	(Additive Identity)
$x = 8$		$x = -8$	(Multiplicative Inverse)

Example 2:

Solve: $|2x - 5| = 11$

Solution:

$2x - 5 = 11$	or	$2x - 5 = -11$	(Definition of Absolute Value)
$2x = 16$		$2x = -6$	(Additive Identity)
$x = 8$		$x = -3$	(Multiplicative Inverse)

Exercise: Answer all the following questions. Write the answer in the space provided.

1. $|2x + 15| = 61$ Answer:________
2. $|x - 7| = 3$ Answer:________
3. $|x - 9| = 6$ Answer:________
4. $|2x + 6| = 10$ Answer:________
5. $|3x + 9| = 12$ Answer:________
6. $|2x + 5| = -8$ Answer:________
7. $|x| = 9$ Answer:________
8. $|x| = 7$ Answer:________
9. $|3x - 7| = -9$ Answer:________
10. $|3x + 7| = 9$ Answer:________

ABSOLUTE VALUE INEQUALITIES

To solve Absolute Value inequalities, the following is used:

For any variable expression E and any non-negative number K, the following is true:

$$|E| \leq K \text{ if and only if } -K \leq E \leq K$$
$$|E| \geq K \text{ if and only if } E \leq -K \text{ or } E \geq K$$

Example 1:

$|3 - 4x| < 11$		(Property of Absolute Value)
$-11 < 3 - 4x < 11$	Add a -3 to both sides	(Additive Identity)
$-14 < -4x < 8$		
$-2 < x < \frac{7}{2}$	Multiply each side by $-\frac{1}{4}$	(Multiplicative Inverse)

Example 2:

$|2x - 4| \geq 8$

$2x - 4 \geq 8$ or $2x - 4 \leq -8$ (Property of Absolute Value)

$2x \geq 12$ or $2x \leq -4$ Add a +4 to both sides (Additive Identity)

$x \geq 6$ or $x \leq -2$ Multiply each side by ½ (Multiplicative Inverse)

Exercise: Answer all of the following questions: Fill in the correct answer in the space provided.

1. $|3x - 1| > 4$ Answer:________
2. $|x + 6| \geq 7$ Answer:________
3. $|3x - 10| \leq 14$ Answer:________
4. $|5x - 4| > 20$ Answer:________
5. $|x - 7| \leq 12$ Answer:________
6. $|3x + 8| \leq 9$ Answer:________
7. $|x + 3| \leq 5$ Answer:________
8. $|4x - 2| > 6$ Answer:________
9. $|x + 7| \geq 9$ Answer:________
10. $|3x - 12| \leq 16$ Answer:________
11. $|6x - 7| \geq 18$ Answer:________
12. $|x - 9| \leq 12$ Answer:________
13. $|7x - 8| \geq 9$ Answer:________
14. $|3x - 6| \leq 12$ Answer:________
15. $|2x + 9| \geq 18$ Answer:________

SECTION 22

RELATIONS AND FUNCTIONS

relation is defined as a set of ordered pairs. A set of all first members of the ordered pairs is called the domain. The second members of the ordered pairs of a relation is called the range.

A function is a relation where no two ordered pairs have the same first member.

For example: Does the ordered pairs {(1,2) (3,3) (4,3) (5,9) (6,7)} define a function?

The Domain: $x = \{ 1, 3, 4, 5, 6 \}$

The Range: $x = \{ 2, 3, 7, 9 \}$

These ordered pairs define a function as no two ordered pairs have the same first value.

FUNCTIONAL NOTATION

Functions are named by using a letter or a combination of letters such as f or g. f(x) is read "f of x" or "the value of f at x." Finding the value of f(x) is referred to as evaluating f at x. To evaluate f(x) at $x = b$, substitute b for x and simplify.

Properties of Functions

1. If f and g are two functions with the same domain, then the sum is written as: $(f + g)(x) = f(x) + g(x)$.
2. The difference of f and g is defined by: $(f - g)(x) = f(x) - g(x)$.
3. The quotient of f and g is defined by: $\left(\frac{f}{g}\right)(x) = \frac{f(x)}{g(x)}$, where $g(x) \neq 0$.
4. The product of f and g is defined by: $(fg)(x) = f(x) \times g(x)$.

Example 1:

If $f(x) = x^2 - 7$ evaluate each of the following:

a. $f(-6)$ b. $f(2c)$ c. $2f(c)$ d. $f(a + 6)$

a. $f(x) = x^2 - 7$

$f(-6) = (-6)^2 - 7$ — Substitute -6 for x

$f(-6) = 36 - 7$ — Order of Operations

$f(-6) = 29$

b. $f(x) = x^2 - 7$

$f(2c) = f(2c)^2 - 7$ — Substitute 2c for x

$f(2c) = 4c^2 - 7$ — Order of Operations

c. $f(x) = x^2 - 7$

$2f(c) = 2(c^2 - 7)$ — Substitute 2 times the function and substitute c for x

$2f(c) = 2c^2 - 14$

d. $f(x) = x^2 - 7$

$f(a + 6) = (a + 6)^2 - 7$ — Substitute $a = 6$ for x

$f(a + 6) = a^2 + 12a + 36 - 7$ — Order of Operations

$f(a + 6) = a^2 + 12a + 29$ — Simplify

Example 2:

If $f(x) = x^2 - 3x + 7$, $g(x)\ 3x^2 - 7$, and $h(x) = f(x) + g(x)$, find $h(4)$.

Solution:

$h(x) = f(x) + g(x)$ — Substitute

$h(x) = (x^2 - 3x + 7) + (3x^2 - 7)$ — $f(x) = x^2 - 3x + 7, g(X) = 3x^2 - 7$

$h(x) = 4x^2 - 3x$ — Simplify

$h(x) = 4(4)^2 - 3(4)$ — Substitute$X = 4$

$h(x) = 4(16) - 3(4)$ — Order of Operations

$h(4) = 64 - 12$

$h(4) = 52$

Exercise: Evaluate each of the following functions. Write the correct answer in the space provided.

1. If $f(x) = 3x - 7$, find:

a. $f(3)$ Answer: ______________

b. $f(-2)$ Answer: ______________

c. $f(0)$ Answer: ______________

d. $f(h + 2)$ Answer: ______________

e. $f(c)$ Answer: ______________

f. $f(c - 1)$ Answer: ______________

2. If $g(x) = 3x^2 + 6$, find:

   a. $g(-3)$ Answer: ____________
   b. $g(6)$ Answer: ____________
   c. $g(0)$ Answer: ____________
   d. $g(h)$ Answer: ____________
   e. $g(h - 1)$ Answer: ____________
   f. $g(c + 5)$ Answer: ____________

3. If $A(x) = 3x^2 + 7x - 9$, find:

   a. $A(-3)$ Answer: ____________
   b. $A(Y)$ Answer: ____________
   c. $A(0)$ Answer: ____________
   d. $A(-8)$ Answer: ____________
   e. $A(Y - 2)$ Answer: ____________
   f. $A(Y + H)$ Answer: ____________

4. If $g(x) = x^2 - 9x + 7$ and $f(x) = -x^2 - 7$, and $h(x) = f(x) + g(x)$, find $h(7)$.
   Answer: ____________

5. If $g(x) = -3x^2 + 9$ and $f(x) = x^2 - 7$, and $h(x) = g(x) - f(x)$, find $h(-9)$.
   Answer: ____________

6. If $f(x) = x^3 + 3x - k$ and $f(2) = 10$, then $k = ?$
   Answer: ____________

SECTION 23

GRAPH OF A STRAIGHT LINE

A graph is a picture used to show some information the center of the graph is the origin. At the origin, the center of the graph, both x and y values are zero. The horizontal line represents the x-axis, and the vertical line represents the y-axis.

To draw a graph of a straight line, one must find points that are solutions to the equations. For example, in the equation $x = y + 1$, one must assume values of one of the variables and solve for the other variable. If one assumes $y = 1$, substitute the value in the equation and solve for x; $(x = 2)$. Choose several values for y and find the values for x. Locate them on the graph. The points should connect to a straight line.

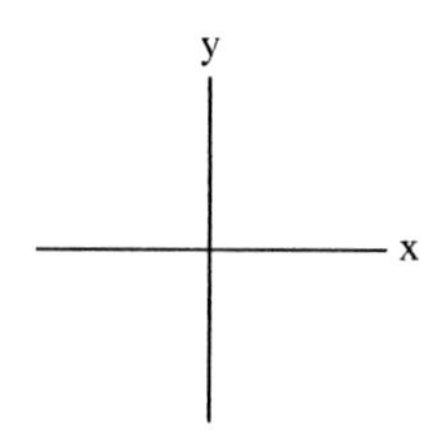

Rules to Graph:

1. Choose at least three different values for one variable.
2. Substitute in the equation and find the value of the other variable.
3. Plot the (x, y) values for each point.
4. Connect the points as a straight line.

In writing a point, use the following notation: (x, y). The x value is written first, and then the y value separated by a comma and enclosed by a parenthesis. The point $(-3, -5)$ has an x value of -3 and a y value of -5.

To Determine if a Point Lies on a Graph

Rules:

1. Substitute in the equation the values given.
2. If both sides of the equation have the same value, the points is a solution to the equation.

Example 1: Which of these points lies on the graph $4x + y = 6$?

a) (2, 7) b) (−3, −5)
c) (1, 2) d) (6, 7)

Solution: Solve equation by substituting the value of each point for x and y in the equation. Start with choice (a) until you find the correct answer.

1. Substitute (2, 7) 2 for x and 7 for y.

$$4x + y = 6$$
$$4(2) + 7 = 6$$
$$8 + 7 = 6$$ **False**

2. Substitute (−3 −5) −3 or x and −5 for y.

$$4x + y = 6$$
$$4(-3) + (-5) = 6$$
$$-12 - 5 = 6$$ **False**

3. Substitute (1, 2) 1 for x and 2 for y.

$$4x + y = 6$$
$$4(1) + 2 = 6$$
$$4 + 2 = 6$$ **True**

Fill in the letter of the correct answer in the space provided.

1. Which one of the following lies on the graph: $y = 3x + 17$?
   a) (3, 4) b) (−2, −7)
   c) (−3, −4) d) (2, 7) Answer:_________
2. Which one of the following lies on the graph: $2x + 3y = 5$?
   a) (3, 4) b) (1, 1)
   c) (−3, −4) d) (2, 9) Answer:_________
3. Which one of the following lies on the graph: $y = x - 2$?
   a) (3, 7) b) (2, 1)
   c) (4, 2) d) (7, 3)
   e) (7, 4) Answer:_________
4. Which one of the following lies on the graph: $6x - 5y = 1$?
   a) (2, 4) b) (1, 1)
   c) (6, 2) d) (3, 4) Answer:_________

5. Which one of the following lies on the graph: $x - y = 1$

a) (5, 7) b) (4, 6)
c) (3, 2) d) (−4, −3) Answer:________

6. Which one of the following lies on the graph: $y = 4 - x$

a) (3, 4) b) (4, 0)
c) (−2, 4) d) (−4, −5) Answer:________

7. Which one of the following lies on the graph: $5x + y = 0$

a) (3, 4) b) (1, −5)
c) (−2, 4) d) (−4, −3) Answer:________

Plotting Points on the Graph

The figure below is a representation of the Cartesian Plane, so named after the mathematician Des Cartes. The x and y axes are divided into equal units, similar to the real number line.

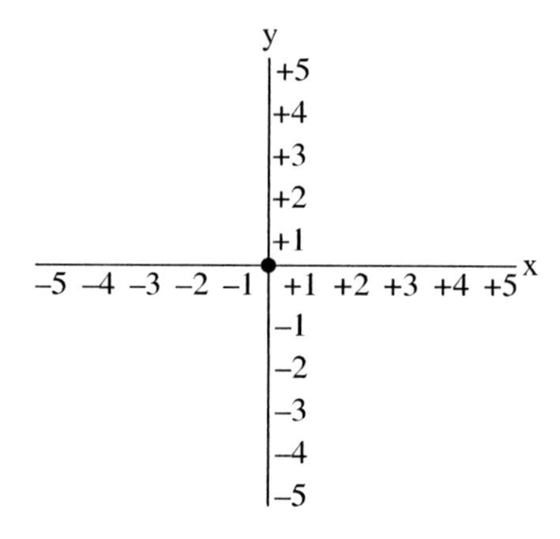

Numbers to the **right** of zero on the x axis are **positive**. Numbers to the **left** of zero on the x axis are **negative**.

Numbers which are on the y axis and **above** the x axis are **positive**. Numbers which are on the y axis and **below** the x axis are **negative**.

Note: If you move right or left but NOT up or down, the point is on the x axis; if you move up or down but NOT right or left, the point is on the y axis.

> Points on the Cartesian Plane are represented by ordered pairs of numbers, sp called because the FIRST number represents the X coordinate and the SECOND number represents the Y coordinate.

The pairs of numbers are written in parentheses and separated by commas: (x, y), (3, −1), (0, 6), etc.

The sign of each number indicates “direction”; that is

if x is positive, the point is to the right of zero

if x is negative, the point is to the left of zero

if y is positive, the point is above zero

if y is negative, the point is below zero.

Example: Plot (locate) the following points in the Cartesian Plane:

$A = (3, 5)$ $B = (-3, 5)$ $C = (-3, -5)$ $D = (3, -5)$

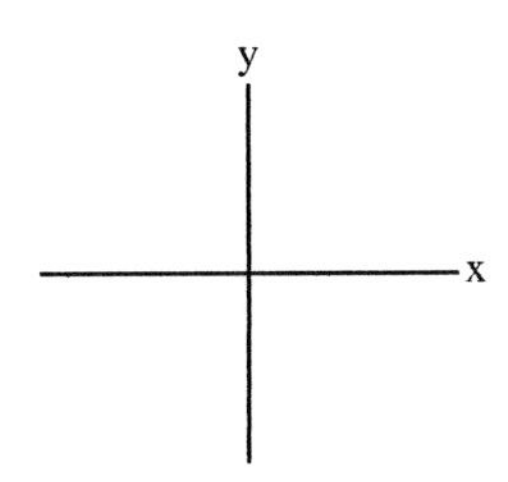

To Find "x" Intercept and the "y" Intercept

Rules Used to find Intercepts:
x—Intercept: Is the point where the line crosses the x axis. At this point y = 0.
y—Intercept: Is the point where the line crosses the y axis. At this point x = 0.

1. To find the x intercept, set the y value equal to to zero.

   Example: $2x + 3y = 6$
   Solution: $2x + 3y = 6$
   $y = 0$
   Then, $2x = 6$
   $x = 3$
   Answer: $x = 3$

2. To find the y intercept, set the value of x equal to zero.

   Example: $3x + 7y = 14$
   Solution: $3x + 7y = 14$
   $x = 0$
   then, $7y = 14$
   $y = 2$
   Answer: $y = 2$

Fill in the correct answer in the space provided

1. Find the x intercept for the equation: $2x - y = 6$. Answer:__________
2. Find the x intercept for the equation: $-2x - 3y = -6$ Answer:__________
3. Find the x intercept for the equation: $-7x - 7y = 14$ Answer:__________
4. Find the x intercept for the equation: $3x + 6y = 12$ Answer:__________
5. Where does the graph $2x - 3y = 6$ intercept the x axis? Answer:__________
6. Where does the graph $-2x - 7y = 14$ intercept the y axis? Answer:__________
7. Where does the graph $3x + 6y = 1$ intercept the x axis? Answer:__________
8. Where does the graph $-4x - 6y = -3$ intercept the y axis? Answer:__________

Write an equation of a Straight Line knowing The "X" intercept and the "Y" intercept

1. Use the formula:

   $\frac{x}{a}+\frac{y}{b}=1$, where "a" is the value of the "x" intercept and "b" is the value of the "y" intercept.

2. Remove the fraction by using the lowest Common Denominator(LCD)

Example 1: Write an equation for the line

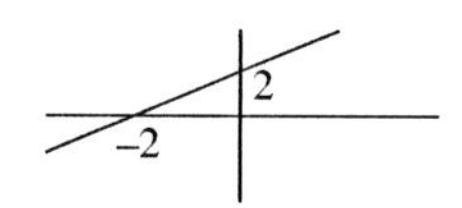

Solution: $\frac{x}{a}+\frac{y}{b}=1$

$\frac{x}{2}+\frac{y}{-2}=1$ Substitute:
a = x − intercept = 2 and
b = y − intercept = −2

$2\left(\frac{2}{2}\right)+2\left(\frac{y}{-2}\right)=2(1)$

x − y = 2 Use LCD = 2 to clear fractions by multiplying both sides by 2

Answer: x − y = 2

Write the correct answer in the space provided:

1. Write an equation for the line

(a)

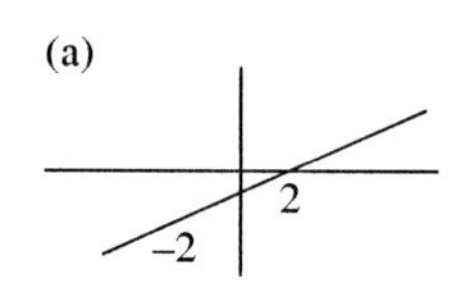

Answer:__________

(b)

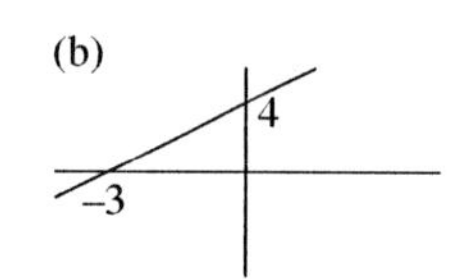

Answer:__________

2. Write an equation for the line

(a)

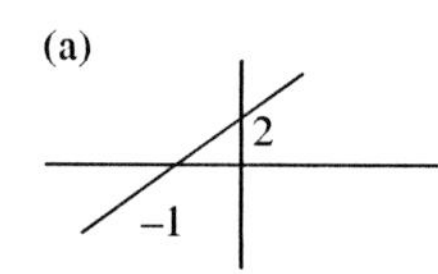

Answer:__________

(b)

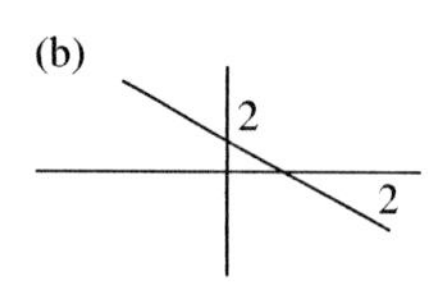

Answer:__________

To Find the Slope given Two Points

Rules:

Slope: It is the ratio of the change in the vertical distance to the change in the horizontal distance

Definition: If points(x_1, y_1) and (x_2, y_2) are any two points, then the slope of the line on which they lie is:

$$\text{Slope} = m = \frac{y_2 - y_1}{x_2 - y_1}$$

1. Label the points (x_1, y_1) and (x_2, y_2)
2. Write the formula for slope.
3. Substitute the values into the formula.
4. Evaluate

Example: Find the slope of the line between the points. (4, 3)(6, 5)

Solution: 1. Label the points

$$(x_1, y_1) = (4, 3)$$
$$(x_2, y_2) = (6, 5)$$

2. Substitute the formula and evaluate.

$$m = \frac{y_2 - y_1}{x_2 - y_1} = \frac{5 - 3}{6 - 4} = \frac{2}{2} = 1$$

Answer: $m = 1$

Find the slope for each set pf points. Write the correct answer in the space provided:

1. (3, 1)(5, 4) Answer:_________
2. (2, 1)(5, 4) Answer:_________
3. (1, 3)(5, 2) Answer:_________
4. (−1, −1)(−2, −2) Answer:_________
5. (1, −3)(4, 2) Answer:_________
6. (2, −4)(3, −1) Answer:_________
7. (4, 5)(5, 6) Answer:_________
8. (−3, −2)(1, 3) Answer:_________
9. (2, −5)(3, −2) Answer:_________
10. (−3, 3)(3, −1) Answer:_________

Slope Intercept Form of the Equation of a Line

The equation of a line "m" and y intercept, "b" is always given by the following:

$y = mx + b$

Rules:

1. Write the equation: $y = mx + b$
2. Substitute in the equation the value of the slope and the value of the y–intecept.

Example: Write the equation of a line with slope 2 and y-intercept 5.

Solution: Substitute $m = 2$ and $b = 5$ in the following equation:

$$y = mx + b$$
$$y = 2x + 5$$

Answer:: $y = 2x + 5$

For each of the following problems, write an equation:
Write the correct answer in the space provided:

1. $m = 4, b = 1$ Answer:________
2. $m = 3, b = -6$ Answer:________
3. $m = -2, b = 4$ Answer:________
4. $m = -5, b = -6$ Answer:________
5. $m = -1, b = -1$ Answer:________
6. $m = 2, b = 2$ Answer:________
7. $m = \frac{2}{3}, b = \frac{3}{4}$ Answer:________
8. $m = \frac{-2}{5}, b = \frac{3}{4}$ Answer:________

To Find the Slope and the "y" Intercept Given an Equation

Rules:

1. Write the formula in the form of $y = mx + b$ (Slope – Intercept form)
2. Solve the equation for y.

Example: Find the slope and the y-intercept for $4x + 2y = 8$

Solution: Write the equation in the form $y = mx + b$ (solve for y)

1. $4x + 2y = 8$ — Original equation
2. $2y = -4x + 8$ — Add a $-4x$ to each side; Additive Inverse
3. $\frac{1}{2}(2y) = \frac{1}{2}(-4x + 8)$ — Multiply by $\frac{1}{2}$; Multiplicative Inverse
4. $y = -2x + 4$
   y-intercept $= b = 4$

Answer: $m = -2, b = 4$

Find the slope and y-intercept of each of the following equations:
Write the correct answer in the space provided:

1. $4x + y = 6$ Answer:________
2. $3x + y = 4$ Answer:________
3. $2x - y = 5$ Answer:________
4. $5x - y = 6$ Answer:________
5. $6x + 3y = 9$ Answer:________
6. $2x + 4y = 6$ Answer:________
7. $x - y = 2$ Answer:________
8. $2x + 3y = 6$ Answer:________

Find the Equation of a Line Given One Point (x_1, y_1) on the Line and the Slope of the Line

Let us now introduce the point-slope form of a liner equation.

When the slope of a line and a point on the line are known, we can use the point slope form to determine the equation of the line. The point-slope form can be obtained by beginning with the slope between any selected point (x, y) and fixed point (x_1, y_1) on a line.

Point-slope form of a Linear Equation

$y - y_1 = m(x - x_1)$, where m is the slope of the line and (x_1, y_1) is a point on the line.

Example: Write an equation of the line that goes through the point (2, 3) and has a slope of 4.

Solution: the slope m is 4. The point on the line is (2, 3); use this point for (x_1, y_1) in the formula. Substitute 4 for m, 2 for x_1, and 3 for y_1 in the point-slope form of a linear equation.

$y - y_1 = m(x - x_1)$
$y - 3 = 4(x - 2)$
$y - 3 = 4x - 8$
$y = 4x - 5$

The graph of $y = 4x - 5$ has a slope of 4 and passes through the point (2, 3)

Find the equation of the line with slope m and going through the point (x_1, y_1):

1. $m = 3$; point (3, 4) Answer ________
2. $m = -2$; point (−2, 4) Answer: ________
3. $m = 5$; point (−1, −3) Answer: ________
4. $m = -1$; point (5, −2) Answer: ________
5. $m = -6$; point (3, −4) Answer: ________

To Graph a Linear Equation

Assume various values for one of the variables, either x or y and substitute that value in the equation to determine the other variable. Any two ordered pairs (x, y) will determine a solution to the equation by graphing. However, assume at least three values to find three ordered pairs to be graphed. If one of the three pairs does not connect to form a straight line, there is an error in the solution of one of the ordered pairs. Go back and solve all three again to determine the error.

Example: Graph $y = 3x - 4$

Assume the following values for x(0, 1, 3)

Steps Substitute each assumed value in the equation to determine the value of y in order to find three ordered pairs.

Assume: $x = 0$
$y = 3x - 4$
$y = 3(0) - 4$
$y = -4$

substitution solution: $(0, -4)$

Assume: $x = 1$
$y = 3x - 4$
$y = 3(1) - 4$
$y = -1$

substitution solution: $(1, -1)$

Assume: $x = 3$
$y = 3x - 4$
$y = 3(3) - 4$
$y = 9 - 4$
$y = 5$

substitution solution: $(3, 5)$
Plot the following points on the graph: $(0, -4)$, $(1, -1)$, $(3, 5)$

Exercise: On graph paper graph each of the following linear equation. Graph paper can be found in the back of the book

1. $y = 2x - 7$
2. $y = 3x - 9$
3. $2y = 4x + 6$
4. $3y = -3x - 9$
5. $2y + x = 6$
6. $3x - y = 9$
7. $y = 3x - 8$
8. $x - y = 6$
9. $3x - y = 9$
10. $-x -y = 6$
11. $x - y = 7$
12. $2x - 4y = 6$
13. $3x + 6y = 9$
14. $2x - y = 4$
15. $y - 3x = 9$

SECTION 24

Solving Simultaneous Equations – Algebraic Solutions – Graphic Solutions

Simultaneous equations are two equations with two different variables

Rules: Category #1

1. If the equation is in the following form:

$$\begin{aligned} x + y &= 6 \\ x - y &= 2 \end{aligned}$$

   Add both equations to eliminate one of the variables. Then solve for the other variable.

2. Substitute the value found for the variable in either equation and solve the equation for the other variable.

Example1: Solve the system of equations $x + y = 6$ and $x - y = 2$.

Solution:

$x + y = 6$	
$x - y = 2$	
$2x \quad = 8$	Eliminate the +y and the −y in both equations by addition
$(½)2x = ½(8)$	Solve for x by using the multiplicative inverse
$x = 4$	
$4 + y = 6$	
$4 - 4 + y = 6 - 4$	Substitute 4 for x in either equation to find y.
$y = 2$	Additive Inverse of +4 is −4.

Answer: $x = 4$, $y = 2$

Hint: You are given 2 equations with 2 unknowns (x and y). What is your **goal?**
You want to find out this equation:

For what value(s) of x and y satisfy both equations? When you learn about graphing linear equations, you will be able to see graphically what you are solving. What you will be **seeing** on the graph is where these 2 lines intersect. The **intersection** point is the value for x and y that satisfies these 2 equations.

Solve each of the following simultaneous equations.
Fill in the correct answer in the space provided.

1. $x - y = 2$
   $x + y = 4$ Answer:_________

2. $a - b = 5$
   $a + b = 3$ Answer:_________

3. $x - 2b = 6$
   $x + 2b = 3$ Answer:_________

4. $-2x + y = 6$
   $2x + 2y = 6$ Answer:_________

5. $a - b = 6$
   $a + b = 2$ Answer:_________

Rules Category #2

1. If the equation is in the following form:

$$2x - y = 6$$
$$x + 3y = 3$$

   Solve by the following procedure:

2. Multiply each and every number on both sides of the first equation by 3 or multiply the second equation by -2.
3. Add both equations to eliminate one set of variables and solve for the other variable.
4. Substitute the value found for the variable in one of the equations and solve for other variable.

Example 1: Solve the system of equations $2x - y = 6$ and $x + 3y = 3$.

Solution: $2x - y = 6$

$x + 3y = 3$

$6x - 3y = 18$ Multiply top equation on both sides by 3

$x + 3y = 3$ Eliminate the $-3y$ and the $+3y$ in both equations by addition.

$7x = 21$

$1/7 \bullet 7x = 1/7 \bullet 21$ Multiplicative inverse of 7 is 1/7.

$x = 3$

$3 + 3y = 3$ Substitute 3 for x in to find y
$3 - 3 + 3y = 3 - 3$ Additive Inverse of +3 is −3.

$3y = 0$

$1/3 \bullet 3y = 1/3 \bullet 0$ Multiplicative Inverse of 3 is 1/3.

Answer: $x = 3, y = 0$.

Solve each of the following simultaneous equations. Fill in the correct answer in the space provided:

1. $2x - y = 8$
   $x + 3y = 4$ Answer:_________

2. $2x - 6y = 4$
   $x + 2y = 3$ Answer:_________

3. $4x - y = 6$
   $x + 2y = 15$ Answer:_________

4. $2a - 3b = -3$
   $a + 4b = 4$ Answer:_________

5. $3x - 2y = -8$
   $x + 4y = 2$ Answer:_________

Rules: Category #3

1. Some simultaneous equations can be solved by substitution.

$$2x + 3y = 7$$
$$y = x - 1$$

In the above problem, substitute "x − 1" for "y" in the top equation, and then solve for x.

$$2x + 3(x - 1) = 7$$

2. Substitute the value found for the variable in either one of the equations, and solve for the other variable.

Example 1: Solve the system of equations $2x + 3y = 7$ and $y = x - 1$.

Solution: $2x + 3y = 7$ Substitute x −1 for y in the top equation.

$y = x - 1$

$2x + 3(x - 1) = 7$

$2x + 3x - 3 = 7$ Distributive law of multiplication

$5x - 3 = 7$ Addition of like terms

$5x - 3 + 3 = 7 + 3$ Additive Inverse of −3 is +3.

$5x = 10$

$1/5 \bullet 5x = 1/5 \bullet 10$ Multiplicative inverse of 5 is 1/5.

$x = 2$

$y = 2 - 1$

$y = 1$

Substitute 2 for x in the bottom equation to find y.

Answer: $x = 2, y = 1.$

Solve each of the follow simultaneous equations. Fill in the correct answer in the space provided:

1. $3x + 4y = 6$
   $y = x - 1$ Answer:________

2. $4x + 2y = 6$
   $x = y$ Answer:________

3. $2x + 3y = 7$
   $y = x + 4$ Answer:________

4. $2x + 4y = 6$
   $x = y + 3$ Answer:________

5. $3x + 6y = 6$
   $x = 2y + 6$ Answer:________

6. $4x + 6y = 16$
   $y = 2x + 8$ Answer:________

7. $2x + 6y = 6$
   $y = x$ Answer:________

Problem Set #17

Solve each of the following simultaneous equations:

1. $x - y = 7$
   $x + y = 1$

2. $x + 3y = 4$
   $2x - y = 8$

3. $2x + 4y = 6$
   $x = y + 3$

4. $-3x + 2y = 6$
   $3x + y = 3$

5. $A + 2B = 7$
   $A - 3B = 1$

6. $3A - 4B = 0$
   $A = 7 - B$

7. $3A + B = 10$
   $2A + 3B = 9$

8. $A + B = 6$
   $2A - B = 3$

9. $2x + 3y = 9$
   $3x + y = 10$

10. $2A - B = 6$
    $3A + B = 1$

11. $3x + y = 6$
    $y = x + 2$

12. $4A + 3B = 2$
    $A + B = 0$

13. $2A - B = 7$
    $A + B = 2$

14. $3x - y = 4$
    $y = x$

15. $2x + 5y = 9$
    $4x - 3y = 5$

16. $x - 2b = 10$
    $x + 2b = 6$

17. $4A - B = 12$
    $A + 2B = 30$

18. $4x + 8y = 12$
    $x = y$

19. $-4x + 2y = 12$
    $4x + 4y = 12$

20. $2A + 3B = 7$
    $A = B - 4$

21. $4x + 12y = 8$
    $2x + 4y = 6$

Graphic solution of simultaneous equations

Simultaneous equation is a set of two linear equations with two variables. Besides algebraic solution, these equations can be sold by graphing.

GRAPH BOTH EQUATIONS. THE POINT OF INTERSECTION OF THE LINES DRAWN IS A SIMULTANEOUS SOLUTION FOR THE EQUATIONS. ITS COORDINATES ARE THE SAME AS IF THE EQUATIONS WERE SOLVED ALGEBRAICALLY.

Example: Solve the following set of equations by finding the point of intersection on the graph:

$$x + y = 6$$
$$x = 2y = 6$$

For the equation $x + y = 6$,
Assume the following values for x(0, 2, 5):

Substitute each assumed values in the equation to determine the value of y in order to find three ordered pairs.

Assume: $x = 0$ — Equation substitute solve

$x + y = 6$

$y = 6$ — (0, 6)

Assume: $x = 1$ — Equation substitute solve

$x + y = 6$

$1 = y = 6$

$y = 5$ — (1, 5)

Assume: $x = 5$ — Equation substitute solve

$x + y = 6$

$5 + y = 6$

$y = 1$ — (5, 1)

Plot the following points on the graph:

(0, 6), (1, 5), (5, 1)

For the equation $x - 2y = 6$,
Assume the following values for x(0, 2, 4):
Substitute each of the assumed values in the equation to determine the value of y in order to find three ordered paris.

Assume: $x = 0$ — Equation substitute solve

$x - 2y = 6$

$0 - 2y = 6$

$y = -3$ — (0, −3)

Assume: $x = 2$ — Equation substitute solve

$x - 2y = 6$

$2 - 2y = 6$

$y = -2$ — (2, −2)

Assume: $x = 4$ — Equation substitute solve

$x - 2y = 6$

$4 - 2y = 6$

$y = -1$ — (4, −1)

Plot the following points on the graph (0, −3), (2, −2), (4, −1): The point of intersection. The points of both equations is the solution.

Graph: $x + y = 6$ (ordered pairs) and $x - 2y = 6$ (ordered pairs)

(0, 6)	(0, −3)
(1, 5)	(2, −2)
(5, 1)	(4, −1)

Graph on page

Solution set:

Answer: (6, 0)

Exercise: Graph the following sets of simultaneous equations: Write the solution set as an ordered pair:

1. $x + y = 6$
   $x - y = 2$ Answer:________

2. $x - y = 2$
   $x + y = 4$ Answer:________

3. $-2x + y = 6$
   $2x + 2y = 6$ Answer:________

4. $2x - y = 8$
   $x + 3y = 4$ Answer:________

5. $4x - y = 6$
   $x + 2y = 15$ Answer:________

6. $3x - 2y = -8$
   $x + 4y = 2$ Answer:________

7. $x + y = 7$
   $x + y = -1$ Answer:________

8. $2x + 3y = 9$
   $3x + y = 10$ Answer:________

9. $2x - y = 6$
   $3x + y = 1$ Answer:________

10. $3x + y = 6$
    $y = x + 2$ Answer:________

11. $4x + 2y = 6$
    $x = y$ Answer:________

12. $2x + 4y = 6$
    $x = y + 3$ Answer:________

Parallel and Perpendicular Lines:

Consistent system of simultaneous equations are equations that have a solution. Inconsistent system of simultaneous do not have a solution. The straight lines of these equations must be parallel. They will never intersect.

Two non-intersecting lines in a plane are parallel. All horizontal lines are parallel to each other. All veritical lines are parallel to each other.

Two lines that intersect and form adjacent angles, each of which is 90°, are perpendicular lines. Vertical and horizontal lines in a plane are perpendicular to each other.

1. Two lines L1 and L2 are parallelIf and only if $m1 = m2$
2. Two lines L1 and L2 are perpendicularIf and only if $m1 = -1/m2$

Exercise:

Which of the following sets of simultaneous equation form parallel or perpendicular lines. Write the word parallel or perpendicular next to the space provided for the answer:

1. $2x + 3y = 6$

   $4x + 6y = 7$ Answer:_________

2. $y = 2x + 3$

   $y = -\frac{1}{2} + 3$ Answer:_________

3. $y = 3x + 1$

   $y = 3x - 4$ Answer:_________

4. $y = 3x - 7$

   $y = -\frac{1}{3}x + 9$ Answer:_________

5. $y = 2x - 7$

   $y = 2x + 9$ Answer:_________

6. $3x + 4y = 12$

   $6x + 8y = 24$ Answer:_________

7. $x + 3y = 6$

   $y = 3x + 9$ Answer:_________

8. $-2x + y = 12$

   $x + 2y = 36$ Answer:_________

SECTION 25

Properties of Exponents

The expression a^n the "a" is the base and "n" is the exponent. The expression 2^4 can be called exponential notation. This is the product of any expression used repeatedly as a factor.

For example:

$6^4 = 6 \bullet 6 \bullet 6 \bullet 6 = 1{,}296$

$-6^4 = -(6 \bullet 6 \bullet 6 \bullet 6) = -1{,}296$

$(-6)^4 = (-6)\,(-6)\,(-6)\,(-6) = 1{,}296$

In the expression -6^4 the base is 6. In the expression $(-6)^4$ the base is -6. Notice the difference of -6^4 and $(-6)^4$.

Zero Power:

Any non-zero real number x, $x^0 = 1$

Examples:

1. $6^0 = 1$
2. $(7/8)^0 = 1$
3. $-3(4)^0 = -3$
4. $(X^2 - 9)^0 = 1$

General Laws of Exponents

Product:	$x^m \bullet x^n = x^{m+n}$	Multiplication same base add exponents
Quotient:	$x^m/x^n = x^{m-n}$	Division same base subtract exponents.
Raising a number to a power	$(x^m)^n = x^{m \bullet n}$	Raising a number to a power multiply exponents.

Examples:

1. $Y^3 \bullet Y^2 \bullet Y = Y^{3+2+1} = Y^6$ Where no exponent is written it is understood to = 1 Multiplication add exponents same base.

2. $a^7b^9/a^2 b^3 = a^{7-2} b^{9-3} = a^5b^6$ Same base Division – Subtract exponents.

3. $(xy^2)^5 = x^{1\bullet5} y^{2\bullet5} = x^5y^{10}$ Raising a number to a power multiply exponents.

Now we can considered negative exponents.
Suppose $X \neq 0$ and n is a natural number, then $x^{-n} = 1/x^n$ or $1/x^{-n} = x^n$

Examples:

1. $4^{-2} = 1/4^2 = 1/16$

2. $1/6^{-3} = 6^3 = 216$

3. $6^{-2}/9^{-1} = 9/6^2 = 9/36 = ¼$

To simplify variable exponential expression involving negative exponents use all the rules already learned:

Examples:

$(3A^2BC^{-4})^3$
$(3^{1\bullet3} A^{2\bullet3} B^{1\bullet3} C^{-4\bullet3})$ Multiply exponents
$(3^3A^6B^3C^{-12})$
$(27A^6B^3C^{-12})$ Simplify $3^3 = 3 \bullet 3 \bullet 3 = 27$

Answer: $\dfrac{27A^6B^3}{C^{12}}$ Apply rule for negative exponents.

Exercise: Evaluate each of the following expression; Write the answer in the space provided.

1. -6^3 Answer:_________
2. $(-6)^3$ Answer:_________
3. $\dfrac{1}{2^5}$ Answer:_________
4. 6^0 Answer:_________
5. -7^0 Answer:_________
6. 5^{-2} Answer:_________
7. $\dfrac{1}{3^{-5}}$ Answer:_________
8. $\dfrac{6^{-3}}{2^{-3}}$ Answer:_________
9. $\dfrac{2^{-3}}{6^{-3}}$ Answer:_________
10. $-3x^0$ Answer:_________
11. $(-3x)^0$ Answer:_________
12. $\dfrac{1}{4^{-3}}$ Answer:_________

13. $\frac{y^0}{5}$ Answer:________

14. $\frac{2^{-3}}{4^{-2}}$ Answer:________

15. $\frac{4^{-2}}{2^{-3}}$ Answer:________

16. $3x^{-4}$ Answer:________

17. $(3x)^{-4}$ Answer:________

18. $\frac{A^3B^4}{A^5B^2}$ Answer:________

19. $3x^{-2}$ Answer:________

20. $\left(\frac{3x^2y^3}{6x^4y^4}\right)^2$ Answer:________

21. $\left(\frac{x^{-2}y}{x^3y^{-4}}\right)$ Answer:________

22. $\left(\frac{12A^3}{16A^4}\right)$ Answer:________

23. $\left(\frac{-48xy^{10}}{-32x^4y^3}\right)$ Answer:________

24. $\left(\frac{2xy^2z^3}{5xy^2}\right)^3$ Answer:________

25. $(A^{-1}B^2)^{-3}(A^2B^{-4})^{-3}$ Answer:________

26. $\left(\frac{x^{-3}y^{-4}}{x^{-2}y}\right)^{-2}$ Answer:________

SECTION 26

FACTORING A POLYNOMIAL

What is factoring?

If a polynomial can be expressed as the product of two or more algebraic expressions, then each expression of the product is called a **factor** of the given polynomial. Let us **try** that again. Thus to **factor a polynomial** is to **find two** or more algebraic expressions whose **product** is the given polynomial.

COMMON FACTORS

Rules for Common Factors

1. Find the largest number that divides evenly into the coefficients.
2. To factor out a variable as a common factor, the variable must appear in each part of the expression. Factor out the variables to the lowest exponent that appears.
3. Divide each segment of the expression by the common factor.

Example 1: Factor completely $6x^2 - 2$

Solution: $6x^2 - 2$

Step 1: Largest number that divides evenly into the coefficients is 2.

Step 2: Divide each segment of the binomial by 2

$$\frac{6x^2}{2} - \frac{2}{2} = 2(3x^2 - 1)$$

Answer: $2(3x^2 - 1)$

Example 2: Factor completely $10a^3 + 15a^2$

Solution: $10a^3 + 15a^2$

Step 1: Largest number that divides evenly into the coefficients is 5

Step 2: Provided the same variable is present in each segment of the binomial, factor it out to the lowest exponent that is present (a^2).

Step 3: Common factor is $5a^2$
$10a^3 + 15^2 = 5a^2 (2a + 3)$

Answer: $5a^2 (2a + 3)$

For each of the following examples write the common factor in the space provided:

1. $3x^2 + 6$ Answer: ________
2. $12a - 24$ Answer: ________
3. $6x^2 - 16$ Answer: ________
4. $x^2 + 6x$ Answer: ________
5. $a^4 + 7a^3$ Answer: ________
6. $6x^2 + 12x$ Answer: ________
7. $3x^2 - 6x$ Answer: ________
8. $9x^2 - 18$ Answer: ________
9. $3x^2y + 6x$ Answer: ________
10. $10x^2y + 20y^2$ Answer: ________
11. $18x^7y - 9x^6y^6$ Answer: ________
12. $6x^3y^7 - 12x^2y^6$ Answer: ________
13. $x^2 - 4x$ Answer: ________
14. $6a^2 - 12a^2b^2$ Answer: ________
15. $6ab - 12a^2b^2$ Answer: ________
16. $7x^2y^4 - 14x14y^3$ Answer: ________
17. $9x^2 + 18y^2$ Answer: ________
18. $7x^2 + 14xy$ Answer: ________

SECTION

27

Factors: The Difference of Two Squares

Rules for Factoring the Difference of Two Squares

1. The first term of the expression is a perfect square and the last term of the expression is a perfect square separated by a negative sign.
2. Factor out the highest common factor.
3. Find the square root of the first term and then find the square root of the second term. One set of factors is separated by a positive sign and the second set of factors by a negative sign.

Example 1: Factor completely $a^2 - 36$

Solution: $a^2 - 36$

Step1: Determine if the expression is the difference of perfect square.

Step2: Factor out common factors first.

Step3: Take the square root of the terms. One factor has a positive sign and the other a negative sign.

Answer: $(a + 6)(a - 6)$

Example 2: Factor completely $x^2 + 36$

Solution: Not factorable – as there are no common factors and it is not a difference of perfect squares (+ sign separate the x^2 and the 36).

Answer: Cannot be factored.

Example 3: Factor completely $16a^2 - 64$.

Solution: $16a^2 - 64$

$\frac{16a^2 - 64}{16\ 16} = 16(a^2 - 4)$ Common Factor

$16a^2 - 4 = 16(a + 2)(a - 2)$ Use method of difference of perfect squares

Answer: $16(a + 2)\ (a - 2)$

Hint: Let us summarize this rule for you. This is what we must look for:

1. Two terms.
2. Both terms are perfect squares.
3. A minus sign in between the two terms.

For each of the following examples, write the factors in the space provided:

1. $x^2 - 25$ Answer________
2. $y^2 - 64$ Answer________
3. $4x^2 + 36$ Answer________
4. $x^2 - 49$ Answer________
5. $9x^2 - 16y^2$ Answer________
6. $49y^2 - 25x^2$ Answer________
7. $3x^2 - 48$ Answer________
8. $2x^2 - 18$ Answer________
9. $9 - x^2$ Answer________
10. $16 - y^2$ Answer________
11. $x^2 - 16$ Answer________
12. $x^6 - 81$ Answer________
13. $x^8 - 81$ Answer________
14. $9x^4 - 81$ Answer________

SECTION

28

TRINOMIAL FACTORS

RULES FOR FACTORING TRINOMIALS IN WHICH THE COEFFICENT OF x^2 IS 1

1. Factor out the highest common factor
2. To factor one must find two numbers whose product is the last term (constant term) and the sum of the same two numbers is the middle term.

Example 1: Factor completely $x^2 + 7x + 10$

Solution: Two numbers are needed whose product is 10 and the sum of the same two numbers is 7. The numbers are 5 and 2.

Answer: $(x + 5)(x + 2)$.

Example 2: Factor completely $x^2 - 5x - 6$.

Solution: Two numbers are needed whose product is 6 and the sum of the dame two numbers with opposite signs is -5. The numbers are -6 and 1.

Answer: $(x - 6)(x + 1)$

Hint: Do you remember the **FOIL Method?** Well, you are using the FOIL Method in **Reverse**. Always check your answer.

Let us look at an example of the Reverse Foil.

To factor the trinomial $x^2 + 5x + 6$

1. The factors of the first term
2. The factors of the last term
3. The sign of the last term, and
4. The sign and coefficient of the middle term.

In the above expression,

1. The factors of the first term are $x \bullet x$.
2. The factors of the last term are either $6 \bullet 1$ or $3 \bullet 2$.

3. The sign of the last term is positive, so (–6) (–1) and (–2) (–3) are also possible factors.
4. The sign of the middle term is positive, and it has coefficient 5.
5. The combination +2 and +3 will add to +5, and at the same time multiply to +6.
6. The result of factoring $x^2 + 5x + 6$ is $(x + 2)(x + 3)$.

To check the answer, multiply:

$$(x + 2)(x + 3) = x^2 + 3x + 2x + 6 = x^2 + 5x + 6$$

The result is the original expression.

Example: Factor $x^2 - 9x + 20$

1. The factor of the first term are x • x.
2. The factor of the last term are 20 • 1 or 10 • 2 or 5 • 4.
3. The sign of the last term is positive so (–20) (–1), (–10) (–2) or (–5) (–4) are also possible factors.
4. The sign of the middle term is negative and has coefficient 9, so we want the combination of –5 and –4 for our factors, since they add to –9 while multiplying to +20.

Therefore in factored form: $x^2 - 9x + 20 = (x - 5)(x - 4)$

Answer: $(x - 5)(x - 4) = x^2 - 4x - 5x + 20 = x^2 - 9x + 20$

For each of the following expressions write the factors in the space provided:

1. $x^2 + 5x + 4$ Answer: ______________
2. $x^2 + 7x + 12$ Answer: ______________
3. $x^2 - 3x + 2$ Answer: ______________
4. $x^2 - 4x - 5$ Answer: ______________
5. $y^2 + y - 12$ Answer: ______________
6. $y^2 - 9y + 20$ Answer: ______________
7. $x^2 + 7x + 6$ Answer: ______________
8. $x^2 + 13x + 42$ Answer: ______________
9. $y^2 + 3y - 18$ Answer: ______________
10. $x^2 + 15x + 56$ Answer: ______________

Hint: Common Factor First

11. $2x^2 + 6x + 4$ Answer: ______________
12. $2y^3 + 4y^2 + 30y$ Answer: ______________
13. $2y^2 + 18y + 28$ Answer: ______________
14. $5x^3 + 45x^2 + 100x$ Answer: ______________

Problem Set #18

Factor completely each of the following:

1. $y^2 + 7y$
2. $x^2 - 36$
3. $4x^2 - 8x$
4. $x^2 + 6x + 5$
5. $y^2 + y - 6$
6. $3A^2 - 48$
7. $x^2 - 144$
8. $A^2 + 6A + 8$
9. $6x - 24x^3$
10. $x^2 - 5x + 6$
11. $6x^3 - 54x$
12. $x^2 - 9x + 20$
13. $x^2 - 7x + 12$
14. $5B^3 + 45B^2 + 100B$
15. $20x^2y - 30x$
16. $49A^2 - 64B^2$
17. $x^2 - 5x + 4$
18. $x^2 + 9x + 20$
19. $2x^2 + 18x + 28$
20. $36x^2 - 81y^2$
21. $18x^6y^2 - 9x^5y^5$
22. $4x^2 - 16y^2$
23. $x^2 + 3x + 2$
24. $x^2 - 15x + 54$
25. $x^6 - 81$
26. $x^8 - 36$
27. $4x^4 - 81$
28. $7x^3y^2 - 14x^5y^2$
29. $x^2 + 4x - 5$
30. $16 - x^2$
31. $196x^2 - 4y^2$
32. $7x^2 + 21x$

33. $6x^2 + 4b$

34. $x^2 - 3x - 10$

35. $x^2 - 2x - 8$

36. $x^2 - 4x - 32$

37. $x^2 + 3x - 54$

38. $x^2 + 13x + 30$

39. $x^2 + 13x - 30$

40. $x^2 - 6x - 40$

SECTION

29

An Additional Factoring Problem

Trinomials Where the First Term is Greater Than 1

In the last section, we looked at the method for factoring trinomials with the first term equal to one. For example, $x^2 - 7x + 10 = (x - 2)(x - 5)$. Now we would like to look at a trinomial where the **first term** $\neq 1$.

Let us look at $2x^2 + 13x + 15$. You can see that the first term is $2x^2$ not x^2. So how do we factor this? We are going to usd **factoring by trial and error**. Recall that factoring is the reverse of multiplying. Consider the product of the following two binomials:

F O I L

$$(2x + 3)(x + 5) = (2x)(x) + (2x)(5) + 3(x) + 3(5)$$
$$= 2x^2 + 10x + 3x + 15$$
$$= 2x^2 + 13x + 15$$

Notice that the product of the first terms of the binomials gives the x-squared term of the trinomial, $2x^2$. Take note that the product of the last terms of the binomials gives the last term, or constant, of the trinomial, +15. Finally, notice that the sum of the products of the outer terms and inner terms of the binomials gives the middle term of the trinomial, +13x. When we factor a trinomial using trial and error, we make use of these important facts. Note that $2x^2 + 13x + 15$ in factored form is $(2x + 3)(x + 5)$.

$$2x^2 + 13x + 15 = (2x + 3)(x + 5)$$

Let us look at another example:

$$2x^2 + 11x + 15$$

The factors of the **first term**, $\mathbf{2x^2}$, are **2x** and **x.**

The factors of the **last term**, **15**, are **15** and **1**, or **5** and **3**. We have to find the correct combination of the **first** and **last term**, so that the middle term is **11x**.

Let us see how it works:

Trinomial	**Possible Factors**	**Product of First Terms**	**Product of Last Terms**	**Sum of the products of Outer and Inner terms**
	$(2x + 15)(x + 1)$	$2x^2$	15	$2x(1) + 15(x) = 17x$
	$(2x + 1)(x + 15)$	$2x^2$	15	$2x(15) + 1(x) = 31x$
$2x^2 + 11x + 15$	$(2x + 5)(x + 3)$	$2x^2$	15	$2x(3) + 5(x) = 11x$
	$(2x + 3)(x + 5)$	$2x^2$	15	$2x(5) + 3(x) = 13x$

Since $(2x + 5)(x + 3)$ yields the correct x term, 11x, The trinomial $2x^2 + 11x + 15$ factors into $(2x + 5)(x + 3)$. $2x^2 + 11x + 15 = (2x + 5)(x + 3)$

Remember: we can check this factoring using the FOIL method.

$$\begin{aligned}\text{Check: } (2x+5)(x+3) &= 2x(x) + 2x(3) + 5(x) + 5(3)\\ &= 2x^2 + 6x + 5x + 15\\ &= 2x^2 + 11x + 15\end{aligned}$$

Since we obtained the original trinomial, our factoring is correct.

Hints:

1. Determine if there is any common factor to all three terms. If so, factor it out.
2. Write all pairs of factors of the coefficient of the squared term(First term)
3. Write all pairs of factors of the constant term(Last term).
4. Try all the combination of these factors until the correct middle term is found.

Factor completely if possible each of the following problems:

1. $3x^2 - 37x + 12$
2. $2x^2 - x - 2$
3. $8x^2 - 2x - 3$
4. $15x^2 - 23x + 4$
5. $6x^2 + 11x - 35$
6. $8x^2 + 11x + 3$
7. $3x^2 - 10x + 3$
8. $6x^2 + 41x - 7$
9. $7y^2 - 40y - 12$

10. $22x^2 + 5x - 3$

11. $12x^2 - 11x + 2$

12. $6x^2 - 7x + 3$

13. $6x^2 - 7x + 6$

14. $10x^2 - x - 9$

15. $2x^2 + 11x + 15$

16. $2x^2 - x - 15$

17. $2x^2 + 5x - 12$

18. $6x^2 - 17x + 12$

19. $2x^2 - 9x + 10$

20. $3x^2 + 2x - 5$

21. $3x^2 + 5x + 2$

22. $3x^2 - 2x - 8$

23. $3x^2 - 11x - 6$

24. $6x^2 + 7x - 10$

25. $8x^2 + 13x - 6$

FACTOR: SUM OR DIFFERENCE OF TWO PERFECT CUBES

$a^3 + b^3 = (a + b)(a^2 - ab + b^2)$

$a^3 - b^3 = (a - b)(a^2 + ab + b^2)$

Example: 1. $27a^3 + b^3 = (3a)^3 + b^3$ Indicate the sum of cubes $27a^3 + b^3 = (3a + b)(9a^2 - 3ab + b^2)$ Factor

2. $a^3 - 125 = (a)^3 - 5^3$ Indicate the difference of cubes $= (a - 5)(a^2 + 5a + 25)$ Factor

1. $b^3 - 8$ Answer:_________

2. $x^3 + 64$ Answer:_________

3. $64x^3 - 27y^3$ Answer:_________

4. $64x^3 + 27y^3$ Answer:_________

5. $125x^3 + 1$ Answer:_________

6. $8a^3 - 64$ Answer:_________

7. $8a^3 + 64$ Answer:_________

8. $125x^3 - 1$ Answer:_________

PROBLEM SET #19

Review Exercise in Factoring

Factor each of the following examples if possible. Write the correct answer in the space provided.

1. $5x^2y - 25x$ Answer:________
2. $6x + 24$ Answer:________
3. $8y^2 + 12y - 40$ Answer:________
4. $12y^2 + 48y$ Answer:________
5. $16x^2y + 32xy^7$ Answer:________
6. $6x^7y^9 + 12x$ Answer:________
7. $x^2 - 9x + 20$ Answer:________
8. $a^2 + 10a + 24$ Answer:________
9. $x^2 - 12x - 28$ Answer:________
10. $x^2 + 12x - 28$ Answer:________
11. $x^2 - 25$ Answer:________
12. $x^2 - 81$ Answer:________
13. $x^3 + 36$ Answer:________
14. $1 - 100y^2$ Answer:________
15. $1 + 9x^2$ Answer:________
16. $16x^2 - 100y^2$ Answer:________
17. $9x^2 - 36$ Answer:________
18. $x^2 - 14x + 49$ Answer:________
19. $x^2 + 14x + 49$ Answer:________
20. $x^2 + 24x + 144$ Answer:________
21. $x^2 - 24x + 144$ Answer:________
22. $4x^2 + 12x + 9$ Answer:________
23. $25A^2 + 40A + 16$ Answer:________
24. $x^4 - y^4$ Answer:________
25. $9A^8 - 16B^9$ Answer:________
26. $27Y^8 - 64X^8$ Answer:________
27. $64X^3 + 27$ Answer:________
28. $64X^3 - 27$ Answer:________
29. $X^3 + 64$ Answer:________
30. $X^3 - 64$ Answer:________

SECTION

30

Rational Expressions

A Rational Expression is an expression in algebra that can be written as the quotient of two Polynomials, $\frac{X}{Y}$, $Y \neq 0$.

To Reduce a Rational Expression

To reduce a rational expression to its simplest form, factor completely both the numerator and denominator. Eliminate the factors that are the same in the numerator and denominator. A rational expression is completely simplified when 1 is the only common factor for both the numerator and denominator.

Example 1: $\frac{X^2 - 9}{X^2 + 7X + 12}$

$= \frac{(X-3)(X+3)}{(X+3)(X+4)}$ Factor

$= \frac{X-3}{X+4}$ After canceling out like factors

Answer: $\frac{X-3}{X+4}$

Example 2: $\frac{X^2 + 3X + 2}{X^2 - 4}$

$= \frac{(X+2)(X+1)}{(X+2)(X-2)}$ Factor

$= \frac{X+1}{X-2}$ Cancel out like factors

Answer: $\frac{X+1}{X-2}$

Exercise: Simplify the following expression. Write the answer in the space provided.

1. $\dfrac{X^2 + 5X - 6}{X^2 - 1}$ Answer:________

2. $\dfrac{X^2 + 7X + 12}{X^2 + 6X + 8}$ Answer:________

3. $\dfrac{X^2 - 9}{X^2 - 6X + 9}$ Answer:________

4. $\dfrac{A^3 - 8}{A^2 - 4}$ Answer:________

5. $\dfrac{X^3 + 125}{X^2 + 10X + 25}$ Answer:________

6. $\dfrac{X^2 - 9}{X^4 - 18X^2 + 81}$ Answer:________

Multiplication and Division of Rational Expressions

1. $\dfrac{X^2 + 8X + 16}{X^2 + X - 6} \bullet \dfrac{X^2 - 4}{X^2 + 6X + 8}$

$= \dfrac{(X + 4)(X + 4)}{(X + 3)(X - 2)} \bullet \dfrac{(X + 2)(X - 2)}{(X + 4)(X + 2)}$ Factor and cancel like factors

$= \dfrac{X + 4}{X + 3}$

Answer: $\dfrac{X + 4}{X + 3}$

2. $\dfrac{Y^2 + 6Y + 9}{Y^2 - 4} \div \dfrac{Y^2 - 9}{Y^2 - 5Y + 6}$

$= \dfrac{Y^2 + 6Y + 9}{Y^2 - 4} \bullet \dfrac{Y^2 - 5Y + 6}{Y^2 - 9}$ Reciprocal (Invert the denominator and change to multiplication).

$= \dfrac{(Y + 3)(Y + 3)}{(Y + 2)(Y - 2)} \bullet \dfrac{(Y - 3)(Y - 2)}{(Y + 3)(Y - 3)}$ Factor and cancel like factors

$= \dfrac{Y + 3}{Y + 2}$ Answer:________

Exercise: Answer each of the following questions. Write the answer in the space provided.

1. $\dfrac{Y^2 - 16}{Y^2 + 7Y + 12} \bullet \dfrac{Y^2 - 4Y - 21}{Y^2 - 4Y}$ Answer:________

2. $\dfrac{X^2 - 81}{X^2 - 16} \div \dfrac{X^2 - X - 20}{X^2 + 5X - 36}$ Answer:________

3. $\dfrac{X^2-36}{X^2-4} \bullet \dfrac{X^2-8X+12}{X^2-12X+36}$ Answer:_________

4. $\dfrac{3X-15}{2X^2-50} \bullet \dfrac{6X+9}{2X^2+5X+3}$ Answer:_________

5. $\dfrac{A^2+A}{2A+3} \bullet \dfrac{3A^2+19A+28}{A^2+5A+4}$ Answer:_________

6. $\dfrac{X^2-9}{X^2-64} \div \dfrac{X^2+4X+3}{X^2+9X+8}$ Answer:_________

To Determine the LCD of Rational Expression

1. Completely factor each denominator. Express repeated factors.
2. Find the largest power of each factor. The LCD is the product of each factor raised to its largest power.

Example:

$$\frac{1}{Y+5} \text{ and } \frac{7}{2Y-9}$$

The LCD for the above rational expressions is:

$$(Y+5)(X-9)^3$$

Look at the following rational expressions:

$$\frac{6X}{(X+6)(X-9)^3} \text{ and } \frac{9}{X(X+6)^2(X-9)}$$

The LCD for the above rational expression is:

$$X(X+6)^2(X-9)^3$$

Addition and Subtraction of Rational Expressions

Example 1: $\dfrac{x}{x^2-9}+\dfrac{2x-1}{x^2+7x+12}$

Factor each denominator to determine LCD.

$$x^2-9=(x+3)(x-3)$$

$$x^2+7x+12=(x+4)(x+3)$$

$$\text{LCD}=(x+3)(x-3)(x+4)$$

$\dfrac{x(x+4)}{(x+3)(x-3)(x+4)}+\dfrac{(2x-1)(x-3)}{(x+3)(x-3)(x+4)}$ Re-write rational expressions using LCD

$$\frac{x^2+4x}{(x+3)(x-3)(x+4)}+\frac{2x^2-7x+3}{(x+3)(x-3)(x+4)}$$ Multiply factors in the numerators

Answer: $$\frac{3x^2-3x+3}{(x+3)(x-3)(x+4)}$$ Combine like terms

Example 2: $$\frac{3x}{x^2-16}-\frac{x-7}{x^2+7x+12}$$

Factor each denominator to determine LCD.

$$x^2-16=(x+4)(x-4)$$

$$x^2+7x+12=(x+4)(x+3)$$

$$\text{LCD}=(x+4)(x-4)(x+3)$$

$$\frac{3x(x+3)}{(x+4)(x-4)(x+3)}-\frac{(x-7)(x-4)}{(x+4)(x-4)(x+3)}$$ Re-write rational expressions using LCD

$$\frac{3x^2+9x}{(x+4)(x-4)(x+3)}-\frac{(x^2-11x+28)}{(x+4)(x-4)(x+3)}$$ Multiply factors in the numerators

$$\frac{3x^2+9x}{(x+4)(x-4)(x+3)}+\frac{-x^2+11x-28}{(x+4)(x-4)(x+3)}$$ Sign number rule for subtraction

Answer: $$\frac{2x^2+20x-28}{(x+4)(x-4)(x+3)}$$ Combine like terms

Exercise: Answer the following questions. Write the answer in the space provided.

1. $\dfrac{y}{y^2-4}-\dfrac{2y-1}{y^2-3y-10}$ Answer:________

2. $\dfrac{4x}{2x-3}+\dfrac{5x}{x-5}$ Answer:________

3. $\dfrac{3x}{x^2-9}-\dfrac{3x-1}{x^2-7x+12}$ Answer:________

4. $\dfrac{6y-7}{y+4}-\dfrac{2y+3}{y^2-16}$ Answer:________

5. $\dfrac{x}{x-5}+\dfrac{7x}{x+3}$ Answer:________

6. $\dfrac{2y-7}{y^2-9}-\dfrac{7y}{y^2+7y+12}$ Answer:________

7. $\dfrac{3x}{x-7}+\dfrac{x^2+9}{x-7}$ Answer:________

8. $\dfrac{3a-7}{a^2+7a+10}+\dfrac{-7a}{a^2+9a+20}$ Answer:________

Review exercise in rational expressions

I. Simplify each of the following. Write the answer in the space provided.

1. $\dfrac{y^2 - y - 20}{3y - 15}$ Answer:_________

2. $\dfrac{x^2 - 4}{x^2 + 5x + 6}$ Answer:_________

3. $\dfrac{x^2 + 3x - 40}{x^2 - 25}$ Answer:_________

4. $\dfrac{3x^2 - 6x}{6x - 12}$ Answer:_________

5. $\dfrac{x^2 - 25}{x^2 + 10x + 25}$ Answer:_________

6. $\dfrac{x^2 + 4x + 3}{x^2 + 6x + 5}$ Answer:_________

7. $\dfrac{2y^2 - 5y - 12}{2y^2 + 5y + 3}$ Answer:_________

8. $\dfrac{x^2 - 9x}{x^2 - 6x - 27}$ Answer:_________

II. Perform the indicated operations on the following rational expressions. Write the answer in the space provided.

1. $\dfrac{x^2 - 36}{x^2 - 5x - 6} \bullet \dfrac{x^2 - 1}{x^2 + 9x + 18}$ Answer:_________

2. $\dfrac{a^2 + 9a}{a^2 - 64} \div \dfrac{a}{a^2 + 16a + 64}$ Answer:_________

3. $\dfrac{x}{x^2 - 9} - \dfrac{3x - 1}{x^2 + 7x + 12}$ Answer:_________

4. $\dfrac{x}{x - 5} + \dfrac{9x}{x + 2}$ Answer:_________

5. $\dfrac{x^2 - 81}{x^2 - 16} \bullet \dfrac{x^2 - x - 20}{x^2 + 5x - 36}$ Answer:_________

6. $\dfrac{3x}{x^2 + 6x} + \dfrac{x - 7}{x^2 - 36}$ Answer:_________

7. $\dfrac{x^2 - 9x}{x^2 - 1} \bullet \dfrac{x^2 - 8x - 9}{x^2 - 18x + 81}$ Answer:_________

8. $\dfrac{3x}{x^2 + 4x + 4} - \dfrac{x - 7}{x^2 - 4}$ Answer:_________

Complex Numbers

SECTION

31

A complex number is an expression in the form of a + bi where "a" and "b" are real numbers and $i = \sqrt{-1}$. In the complex number, "a" is called the real part and bi is called the imaginary part. The conjugate of a complex number a + bi is a − bi.

1. Addition and subtraction of complex numbers – Add or subtract the real and imaginary parts

Example:

1. $(9 - 2i) + (-7 + 5i)$
   Line up like terms:
   $$\begin{array}{r} 9 - 2i \\ \underline{-7 + 5i} \end{array}$$
   Answer: $2 + 3i$

2. $(-7 + 6i) - (2 - 9i)$
   Line up like terms:
   $$\begin{array}{r} -7 + 6i \\ \underline{-2 + 9i} \end{array}$$
   Answer: $-9 + 15i$

Exercise: Perform indicated operations. Write the correct answer in the space provided.

1. $(6 + 2i) + (6 - 9i)$ Answer:________
2. $(4 - 9i) + (3 - 7i)$ Answer:________
3. $(-3 - 5i) - (5 - 8i)$ Answer:________
4. $(1 - 3i) - (7 - 2i)$ Answer:________
5. $(7 + 3i) - (9 - 8i)$ Answer:________
6. $(3 - 7i) + (-7 - 4i)$ Answer:________
7. $(9 - i) + (-9 - i)$ Answer:________
8. $(9 - i) - (-9 - i)$ Answer:________
9. $(6 - 3i) - (-6 + 3i)$ Answer:________
10. $(7 - 12i) + (-9 - i)$ Answer:________

11. $(3 - i) + (-3 + i)$ Answer:________

12. $(3 - i) - (-3 + i)$ Answer:________

13. $(-7 - 12i) - (-9 - 8i)$ Answer:________

14. $(2 - i) + (-2 + i)$ Answer:________

15. $(2 - i) - (-2 + i)$ Answer:________

16. $(7 + 3i) + (-9 - 8i)$ Answer:________

17. $(3 - i) - (-7 + 8i)$ Answer:________

18. $(3 - i) + (-7 + 8i)$ Answer:________

2. To multiply two complex numbers, perform ordinary binomial multiplication and substitute -1 for i^2,

$$i^2 = -1$$

Example:

1. $4i(5i) = 20i^2$ Substitute $i^2 = -1$

 $20(-1)$

 Answer: -20

2. $(3 - 5i)(2 + 6i)$

 $6 + 18i - 10i - 30i^2$ Binomial Multiplication

 $6 + 8i - 30i^2$ Combine Like Terms

 $6 + 8i - 30(-1)$ Substitute $i^2 = -1$

 $36 + 8i$ Simplify

Exercise: Perform the indicated. Write the answer in the space provided.

1. $(4 + 2i)(3 - 4i)$ Answer:________

2. $6i \bullet 9i$ Answer:________

3. $(-6 - 4i)(3 - 7i)$ Answer:________

4. $(-3 - 4i)(2 + 7i)$ Answer:________

5. $(-6i)(-3i)$ Answer:________

6. $(3 - 7i)^2$ Answer:________

7. $(2 - i)^2$ Answer:________

8. $(8 - i)(3 - 7i)$ Answer:________

9. $9i \bullet 7i$ Answer:________

10. $(-7 - 3i)(-2 - 3i)$ Answer:________

11. $(-2 - i)(-3 - i)$ Answer:________

12. $(2 - 7i)(3 - 9i)$ Answer:________

13. $(9i)(3i)$ Answer:________

14. $(-7i)(-9i)$ Answer:__________

15. $(3 - i)(2 - i)$ Answer:__________

16. $(6 - i)^2$ Answer:__________

17. $(7 - 9i)^2$ Answer:__________

18. $(3 - 2i)(6 - 4i)$ Answer:__________

3. Division of complex numbers:

   To divide complex numbers, multiply the numerator and the denominator by the conjugate of the denominator.

   Example: $\dfrac{18 - 9i}{7 + 2i}$

$$\frac{18 - 9i}{7 + 2i} = \frac{18 - 9i}{7 + 2i} \bullet \frac{7 - 2i}{7 - 2i}$$ Multiply numerator and the denominator by conjugate base

$$= \frac{126 - 36i - 63i + 18i^2}{49 - 14i + 14i + 4i^2}$$

$$= \frac{126 - 99i + 18i^2}{49 - 4i^2}$$ Combine like terms

$$= \frac{126 - 99i + 18(-1)}{49 - 4(-1)}$$ Substitute $i^2 = -1$

$$= \frac{126 - 99i - 18}{49 + 4}$$ Simplify

$$= \frac{108 - 99i}{53}$$

Exercise: Perform the indicated operation. Write the answer in the space provided.

1. $\dfrac{2i}{3 + i}$ Answer:__________

2. $\dfrac{4 + 2i}{4 - 2i}$ Answer:__________

3. $\dfrac{-7 + 9i}{4 + i}$ Answer:__________

4. $\dfrac{-7 + 9i}{4 + i}$ Answer:__________

5. $\dfrac{5}{4 + 4i}$ Answer:__________

6. $\dfrac{7}{3 - 9i}$ Answer:__________

7. $\dfrac{7+3i}{2-9i}$ Answer:________

8. $\dfrac{3-i}{3+2i}$ Answer:________

9. $\dfrac{2-i}{7+9i}$ Answer:________

10. $\dfrac{3+i}{2-4i}$ Answer:________

11. $\dfrac{2i}{3+i}$ Answer:________

12. $\dfrac{9-7i}{2-3i}$ Answer:________

13. $\dfrac{3-i}{2+6i}$ Answer:________

14. $\dfrac{7i}{4-i}$ Answer:________

15. $\dfrac{10-i}{20+i}$ Answer:________

Powers of i

$Ei^1 = i$

$Ei^2 = -1$

$Ei^3 = i^2 \bullet i = -1(i) = -i$

$Ei^4 = i^2 \bullet i^2 = (-1)(-1) = +1$

$Ei^5 = i^4 \bullet i = 1 \bullet i = +i$

$Ei^6 = i^4 \bullet i^2 = 1(-1) = -1$

$Ei^7 = i^4 \bullet i^3 = 1(-i) = -i$

$Ei^8 = (i^4)^2 = (1)^2 = 1$

Since $i^4 = 1$, it can be stated that $(i^4)^n = 1^n$ for any integer n. Powers of i can be evaluated by factoring out powers of i^4.

Example:

$i^{17} = i^1 = 1$ The remainder of $17 \div 4 = 1$

Exercise: Perform the indicated. Write the answer in the space provided.

1. i^{19} Answer:________
2. i^{70} Answer:________
3. $-i^{40}$ Answer:________
4. i^{32} Answer:________

5. i^{17} Answer:________

6. i^{40} Answer:________

7. i^{16} Answer:________

8. i^{28} Answer:________

9. i^{9} Answer:________

10. i^{12} Answer:________

Review Exercise: (Complex Numbers)
Perform the indicated. Write the correct answer in the space provided.

Problem Set #21

1. $(-7 + 4i) - (5 - 9i)$ Answer:________
2. $(1 + 3i) + (7 - 9i)$ Answer:________
3. $9i - (2 - 7i)$ Answer:________
4. $(2 + 5i) - (-7 + 9i)$ Answer:________
5. $(3 - i) - (-3 + i)$ Answer:________
6. $(2 + 5i) + (-7 + 9i)$ Answer:________
7. $(-7 - 4i) + (-5 + 9i)$ Answer:________
8. $(3 - i) + (-9 - 7i)$ Answer:________
9. $(2 - i) - (-7 - i)$ Answer:________
10. $(3 + 4i) + (6 - 2i)$ Answer:________
11. $6i \times 8i$ Answer:________
12. $(-7i)(8i)$ Answer:________
13. $(4 - 5i)(4 + 5i)$ Answer:________
14. $(6 + 7i)(6 - 7i)$ Answer:________
15. $(-3 - 4i)(2 + 7i)$ Answer:________
16. $\frac{-7 + 19i}{4 - 3i}$ Answer:________
17. $\frac{1}{7 + 2i}$ Answer:________
18. $\frac{5i}{2 - 4i}$ Answer:________
19. $\frac{-4 + 39i}{5 - 2i}$ Answer:________

20. $(2 + 4i)^2$ Answer:_________

21. $(3 - 7i)^2$ Answer:_________

22. $\frac{6i}{7 - 9i}$ Answer:_________

23. $\frac{2 - 3i}{2 + 3i}$ Answer:_________

24. $-i^{40}$ Answer:_________

25. i^{39} Answer:_________

26. $-i^{10}$ Answer:_________

27. i^{10} Answer:_________

28. $-3i^{18}$ Answer:_________

29. $-4i^{12}$ Answer:_________

30. i^{19} Answer:_________

SECTION

32

SOLVING QUADRATIC EQUATIONS

Rules:

1. Any equation that can be put into the from of $ax^2 + bx + = 0$ is called a quadratic equation. There are two solutions for each equation.
2. Set equation equal to zero
3. Factor.
4. Set each factor equal to zero
5. Solve for each of the factors

Example 1: $x^2 + 7x = -6$

Solution: 1. Set equation equal to zero
$x^2 + 7x + 6 = 0$

2. Factor
$(x + 6)(x+1) = 0$

3. Set each factor equal to zero
$x+6 = 0$ or $x + 1 = 0$

4. Solve
$x = -6$ or $x = -1$

Answer: $(-6, -1)$ solution set

Solve each of the following quadratic equations.

1. $y^2 + 7y + 10 = 0$
2. $x^2 + 12y + 11 = 0$
3. $x^2 + 5x = -4$
4. $y^2 - 9y = -20$
5. $x^2 + 2x = 3$

6. $x^2 = 16$
7. $x^2 - x = 2$
8. $x^2 = 25$
9. $x^2 - 7x = -12$
10. $y^2 + 4y = -3$
11. $x^2 + 5x = -6$
12. $x^2 + 8x = -15$

Problem Set #22

1. $x^2 + 9x = -20$
2. $y^2 - 4y = -3$
3. $x^2 + 9x = -8$
4. $x^2 = 36$
5. $x^2 - 8x = -15$
6. $x^2 - x = 20$
7. $x^2 + 3x = 4$
8. $x^2 + 6x = -8$
9. $x^2 + 7x = -6$
10. $A^2 - 7A = 18$
11. $x^2 + 9x = 10$
12. $A^2 - 7A = 18$
13. $y^2 - 3y = -2$
14. $y^2 - 6y = 16$
15. $x^2 - 9x = -14$
16. $x^2 - 25 = 0$
17. $9x^2 - 36 = 0$
18. $x^2 - 7x = 0$
19. $4x^2 + 5x = 0$
20. $x^2 - 4x - 21 = 0$
21. $2x^2 - 9x + 10 = 0$
22. $x^2 - 9x = 10$
23. $3x^2 + 24x = 0$
24. $x^2 + 8x = -15$

SECTION

33

To Solve Quadratic Equations by using the Quadratic Formula

The Discriminant and the Relationship to the Solutions of a Quadratic Equation:

The equation $ax^2 + bx + c = 0$ containing real coefficients and $a \neq 0$ has its discriminant $b^2 - 4ac$.

- $b^2 - 4ac > 0$ then $ax^2 + bx + c = 0$ has two real solutions
- $b^2 - 4ac = 0$ then $ax^2 + bx + c = 0$ has one real solution. The two solutions are the same.
- $b^2 - 4ac < 0$ then $ax^2 + bx + c = 0$ has two non-real complex solutions. The solutions are conjugates of each other.

Use the following steps to solve a quadratic equation by the formula

(This formula can be used to solve any quadratic equation): $X = \dfrac{-b \pm \sqrt{b^2 - 4ac}}{2a}$

Steps:

1. Set equation = 0
2. Find the values of a, b, and c, "a" represents number in front of x^2 (after setting equation = 0), "b" represents the number and sign of front of x (after setting the equations = 0) "c" represents the constant term (after setting the equation = 0)
3. Write the formula
4. Substitute in the formula
5. Perform order of operations
6. Solve (two solutions)

Solve the following quadratic equations

1. $5x^2 - 5x = 4$

$5x^2 - 5x - 4 = 0$ set equation = 0

$ax^2 + bx + c = 0$ write the value of "a","b","c"

a = 5

b = −5

c = −4

$$x = \frac{-b \pm \sqrt{b^2 - 4ac}}{2a}$$ write the quadratic formula

$$x = \frac{(-5) \pm \sqrt{(-5)^2 - 4(5)(-4)}}{2(5)}$$ substitution

$$x = \frac{5 \pm \sqrt{25 + 80}}{10}$$ order of operations

$$x = \frac{5 \pm \sqrt{105}}{10}$$

The solutions of $5x^2 - 5x = 4$ are

$$\frac{5 + \sqrt{105}}{10} \quad \text{and} \quad \frac{5 - \sqrt{105}}{10}$$

2. $x^2 - 4x = 6$

$x^2 - 4x - 6 = 0$ Set equation = 0

$ax^2 + bx + c = 0$ Write the values of "a", "b" and "c".

a = +1

b = −4

c = −6

$$x = \frac{-b \pm \sqrt{b^2 - 4ac}}{2a}$$ Write the quadratic formula

$$x = \frac{-(-4) \pm \sqrt{(-4)^2 - 4(1)(-6)}}{2(1)}$$ Substitute

$$x = \frac{4 \pm \sqrt{16 + 24}}{2}$$ Order of Operations

$$x = \frac{4 \pm \sqrt{40}}{2}$$

The solutions of $x^2 - 4x = 6$ are:

$$x = \frac{4 \pm \sqrt{40}}{2} \text{ and } x = \frac{4 - \sqrt{40}}{2}$$ Solutions

Exercise: Find the solution set of the following quadratic equations: (**USE ONLY THE QUADRATIC FORMULA**). Write the answer in the space provided.

1 $2y^2 + 4y = -1$ Answer:_________

2. $y^2 - 2y - 15 = 0$ Answer:_________

3. $3x^2 - 5x = 3$ Answer:________
4. $x^2 + 3x = 10$ Answer:________
5. $2x^2 + 5x - 3 = 0$ Answer:________
6. $7y^2 + 3y = 15$ Answer:________
7. $2x^2 + 4x - 1 = 0$ Answer:________
8. $x^2 - 5x - 24 = 0$ Answer:________
9. $24y^2 - 22y = 35$ Answer:________
10. $2x^2 + 8x = -4$ Answer:________
11. $y^2 + 4y = -2$ Answer:________
12. $4x^2 = 12x - 9$ Answer:________
13. $3x^2 = 7x + 6$ Answer:________
14. $5y^2 + 2y = 9$ Answer:________
15. $a^2 + 3a = 5$ Answer:________
16. $12x^2 - 41x = -24$ Answer:________
17. $3x^2 = 7x$ Answer:________
18. $3x^2 - 5x = 9$ Answer:________
19. $5y^2 = -9y$ Answer:________

Examples: For each of the following, state the discriminant and the number of real solutions:

1. $2x^2 - 5x + 2 = 0$

   $b^2 - 4ac$ Discriminant Formula

   $(-5)^2 - 4(2)(2)$ Substitute

   $25 - 16$

   9 Discriminant

   The discriminant is positive. Equation $2x^2 - 5x + 2 = 0$ has 2 real solutions.

2. $4x^2 + 9x + 12 = 0$

   $b^2 - 4ac$ Discriminant Formula

   $(9)^2 - 4(4)(12)$ Substitute

   $81 - 192$

   -111 Discriminant

   The discriminant is negative. Equation $4x^2 + 9x + 12 = 0$ has no real solutions.

3. $x^2 + 4x + 4 = 0$

   $b^2 - 4ac$ Discriminant Formula

   $4^2 - 4(1)(4)$ Substitute

   $16 - 16 = 0$ Discriminant

   The discriminant $= 0$. The equation $x^2 + 4x + 4 = 0$ has one real solution.

Exercise: For each of the following QUADRATIC EQUATIONS, determine the discriminant and state the number of real solutions: Write the answer in the space provided.

1. $3x^2 - x + 10 = 0$ Answer:________
2. $2x^2 - 4x - 7 = 0$ Answer:________
3. $8x^2 = 4x - 3$ Answer:________
4. $x^2 - 20x + 100 = 0$ Answer:________
5. $12x^2 + 15x = -7$ Answer:________
6. $y^2 + 3y + 3 = 0$ Answer:________
7. $x^2 + 8x + 16 = 0$ Answer:________
8. $x^2 - 7x = -9$ Answer:________
9. $x^2 + 3x + 7 = 0$ Answer:________
10. $x^2 + 81x - 17 = 0$ Answer:________

PROBLEM SET #23

Review Exercises in Quadratic Equations

I. Solve each of the following quadratic equations by factoring: Write the answer in the space provided:

1. $x^2 - 7x = 8$ Answer:________
2. $x^2 = -4x - 3$ Answer:________
3. $x^2 + 7x = -10$ Answer:________
4. $x^2 - 6x + 8 = 0$ Answer:________
5. $x^2 + 4x = -4$ Answer:________
6. $x^2 - 6x = -9$ Answer:________
7. $x^2 + 7x = -10$ Answer:________
8. $x^2 - 3x - 4 = 0$ Answer:________
9. $x^2 - 7x = -12$ Answer:________
10. $x^2 + 8x = -16$ Answer:________
11. $y^2 + 12y = -36$ Answer:________
12. $y^2 + 6y = -5$ Answer:________
13. $x^2 + 9x = -14$ Answer:________
14. $x^2 - 8x = -15$ Answer:________
15. $y^2 + 10y = -16$ Answer:________

II. Solve each of the following QUADRATIC EQUATIONS by using the QUADRATIC FORMULA: Write the answer in the space provided.

1. $5y^2 - 5y = 4$ Answer:________
2. $2y^2 - 7y = 9$ Answer:________
3. $x^2 - 4x = 7$ Answer:________
4. $2y^2 + 6x = -4$ Answer:________
5. $24x^2 + 10x = 21$ Answer:________
6. $x^2 + 3x = 11$ Answer:________
7. $2x^2 - 5x = 7$ Answer:________
8. $32x^2 - 44x = -15$ Answer:________
9. $x^2 + 2x = 12$ Answer:________
10. $x^2 - 2x = 15$ Answer:________
11. $3x^2 + 5x = 1$ Answer:________
12. $x^2 = 5x + 24$ Answer:________
13. $3x^2 - 5x = 3$ Answer:________
14. $3x^2 - 5x = 8$ Answer:________
15. $2x^2 + 4x - 1 = 0$ Answer:________

III. For each of the following QUADRATIC EQUATIONS, determine the DISCRIMINANT and state the number of real solutions: Write the answer in the space provided.

1. $4x^2 - x + 10 = 0$ Answer:________
2. $4x^2 - 2x - 7 = 0$ Answer:________
3. $x^2 + 10x + 25 = 0$ Answer:________
4. $x^2 - 8x = -7$ Answer:________
5. $x^2 - 3x = -9$ Answer:________
6. $x^2 + 6x + 1 = 0$ Answer:________
7. $x^2 - 7x - 19 = 0$ Answer:________
8. $y^2 + 4y + 4 = 0$ Answer:________
9. $x^2 - 3x + 6 = 0$ Answer:________
10. $x^2 + 9x = -7$ Answer:________

SECTION 34

Radical Equations

Some equations involving radicals can be solved by using the POWER PRINCIPLE: In the following example, this principle can be used to solve the equation

Example:

$\sqrt{x+5} = 6$	
$\left(\sqrt{x+5}\right)^2 = 6^2$	Square each side of the equation
$x + 5 = 36$	(The square root of the number squared is the number)
$x = 31$	Solution

Exercises: Solve each of the following radical equations. Write the correct answer in the space provided.

1. $\sqrt{x-4} = 6$ Answer:________
2. $\sqrt{10-x} = 4$ Answer:________
3. $x = \sqrt{12x-35}$ Answer:________
4. $2x = \sqrt{4x+15}$ Answer:________
5. $\sqrt{2x+11} - \sqrt{2x-5} = 2$ Answer:________
6. $\sqrt{x+7} + \sqrt{x-5} = 6$ Answer:________
7. $\sqrt[3]{7x-3} = \sqrt[3]{2x+7}$ Answer:________
8. $\sqrt[4]{x^2+20} = \sqrt[4]{9x}$ Answer:________
9. $\sqrt{x+7} - 2 = \sqrt{x-9}$ Answer:________
10. $\sqrt[3]{2x^2+5x-3} = \sqrt[3]{x^2+3}$ Answer:________

UNIT *three*

Section 35: Introduction 121

Section 36: Plane Angle (Degrees – Radians) 122

Section 37: Pythagorean Theorem 124

Section 38: Trigonometric Functions of an Acute Angle 127

Section 39: Trigonometric Functions of 30°, 45°, and 60° 130

Section 40: Values of Trigonometric Functions of Acute Angles other than 30°, 45° and 60° 132

Section 41: Using Table to Evaluate Trigonometric Functions 134

Section 42: To solve for the missing angles and sides of a right triangle 136

Section 43: Reducing Trigonometric Functions To Positive Acute Angles 140

Section 44: Functions of a Negative Angle 142

Section 45: Reference Angles 144

Section 46: Basic Relationship and Identities 147

Section 47: Trigonometric Functions Sum – Difference – Product 150

Section 48: Solving Triangles That Do Not Have a Right Angle — Oblique Triangles 152

SECTION 35

INTRODUCTION

Trigonometry is the division of mathematics that measures the sides and angles of a triangle. Plane trigonometry, which this part of the book is based upon, is restricted to triangle in a plane. Trigonometry deals with the ratios including trigonometric functions. Trigonometric functions are used in engineering and navigation as well as in surveying. They are used to study light, sound, and electricity. In this book we will study the properties and relations of trigonometric functions.

PLANE ANGLE (DEGREES – RADIANS)

SECTION 36

The plane angle, ∠XOY is formed by two rays OX and OY. The point O is called the vertex.

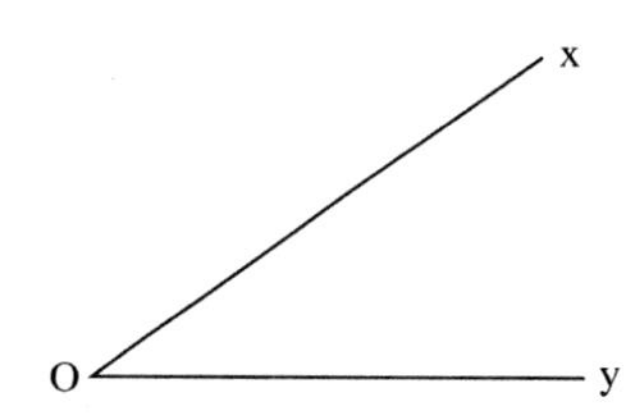

The plane angle has two sides; the initial side OY and the terminal side OX. Angles are measured in units called degrees. A degree (°) is defined as a measure of the arc interrupted by a central angle and is equal to 1/360 of the circumference of a circle. A minute (′) is 1/60 of a degree and a second (″) is 1/60 of a minute or 1/3600 of a degree.

Example 1: Perform the indicated:

(a) $1/4\ (44°\ 16') =$

(b) $1/2\ (91°\ 15')$

(c) $1/3\ (76°\ 15')$

Answers:

(a) $11°\ 4'$

(b) $45°\ 37.5'$ or $45°\ 37'\ 30''$

(c) $25°\ 37.5$ or $25°\ 37'\ 30''$

Instead of using degrees, the central angle of a circle can be measured in terms of *radians*. The circumference of a circle has 2π radians. It can be stated that $2\pi = 360°$.

$$1 \text{ radian} = \frac{180}{\pi} = 57.296° = 57°\ 17'46''$$

$$1 \text{ degree} = \frac{\pi}{180} \text{ radians} = 0.017453 \text{ rad}$$

$$\pi = 3.14159$$

Example 1: Change 7/15 π radians to degrees

Solution: $7/15\pi \bullet \frac{180}{\pi}$ (π in the numerator and π in the denominator cancels out)

$7(180)/15 = 84°$

Answer: 84°

Example 2: Change 60° to radians

Solution: $600 \bullet \frac{\pi}{180}$

$\frac{600\pi}{180} = 3\frac{1}{3}\pi$ radians

Answer: $3\frac{1}{3}\pi$ radians

Exercise: Answer the following questions: Fill in the answer in the space provided.

1. Perform the indicated.

(a) $\frac{1}{3}(66°\ 15')$ Answer:__________

(b) $\frac{2}{3}(9°\ 6')$ Answer:__________

(c) $\frac{1}{2}(61°\ 14')$ Answer:__________

(d) $\frac{1}{2}(71°\ 15')$ Answer:__________

(e) $\frac{2}{3}(12°\ 24')$ Answer:__________

2. Change each of the following radians to degrees. Fill in the correct answer in the space provided.

(a) $\frac{7}{12}\pi$ radians Answer:__________

(b) $\frac{11}{15}\pi$ radians Answer:__________

(c) $\frac{3}{4}\pi$ radians Answer:__________

(d) $\frac{1}{4}\pi$ radians Answer:__________

3. Change each of the following degrees to radians. Fill in the correct answer in the space provided.

(a) 30° Answer:__________

(b) 45° Answer:__________

(c) 20° Answer:__________

(d) 30° Answer:__________

(e) 90° Answer:__________

SECTION 37

PYTHAGOREAN THEOREM

In a right triangle, the sum of the squares of the lengths of the legs is equal to the square of the length of the hypotenuse (the longest side) where "a" and "b" are legs and "c" is the hypotenuse.

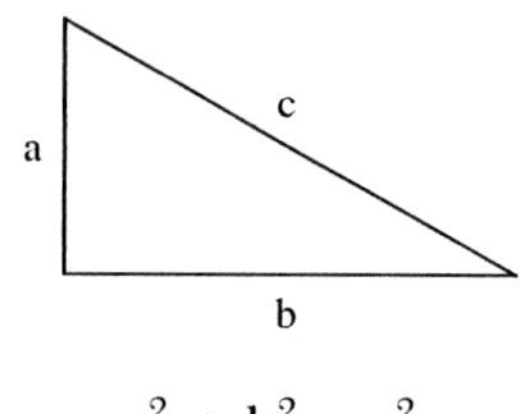

$$a^2 + b^2 = c^2$$

Example 1: In the right triangle, find the value of x

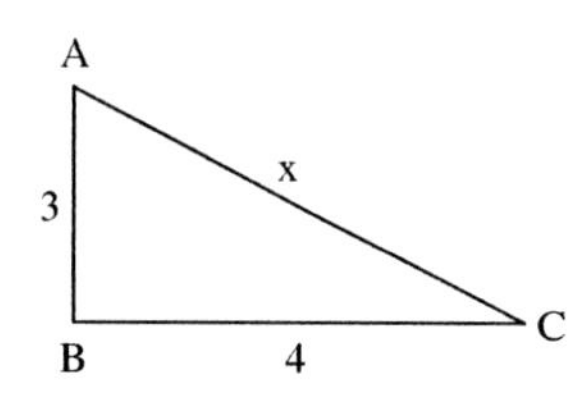

$a^2 + b^2 = c^2$ Pythagorean Theorem

Substitute $4^2 + 3^2 = c^2$

Solve for hypotenuse $16 + 9 = c^2$

$25 = c^2$

$\sqrt{25}$

$c = 5$

Example 2: In the right triangle, find the value of x

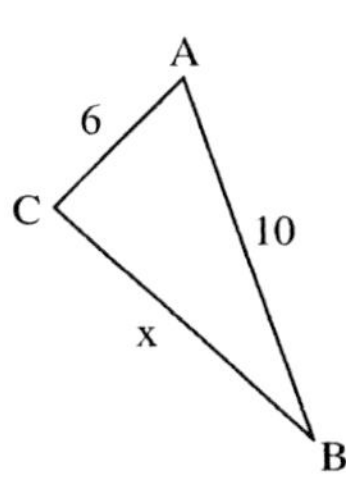

$a^2 + b^2 = c^2$ Pythagorean Theorem

Substitute $x^2 + 6^2 = 10^2$

$x^2 + 36 = 100$

Solve for the missing leg $x^2 + 36 - 36 = 100 - 36$

$x^2 = 64$

$x = \sqrt{64}$

$x = 8$

Exercise: In each of the following right triangles, write the answer in the space provided.

1.

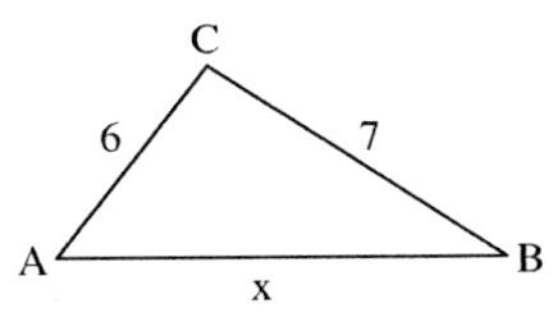

Answer:_________

2.

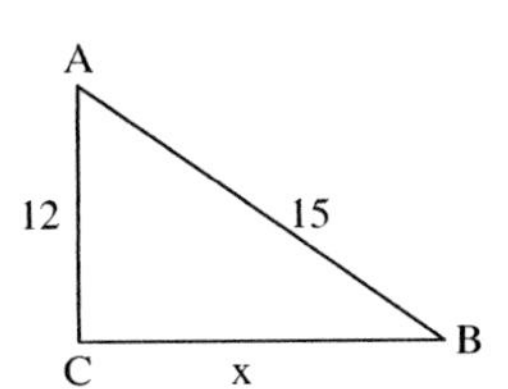

Answer:_________

3.

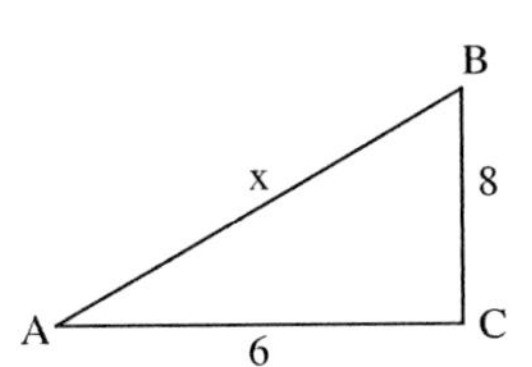

Answer:_________

4.

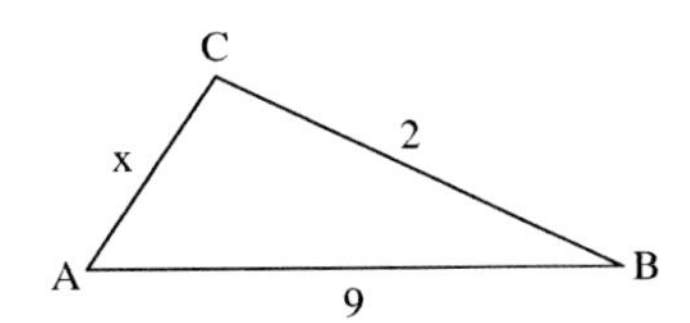

Answer:_________

5.

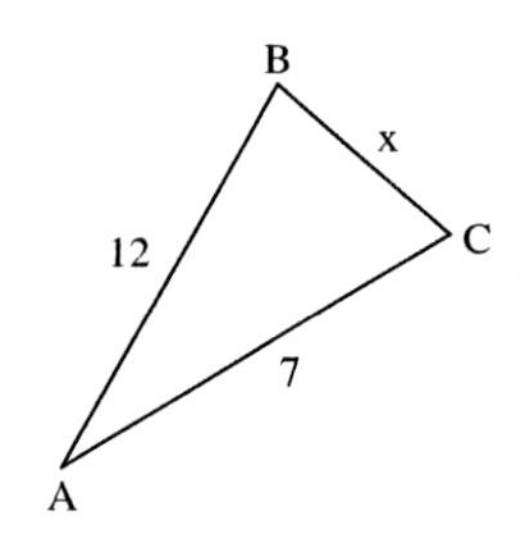

Answer:__________

6.

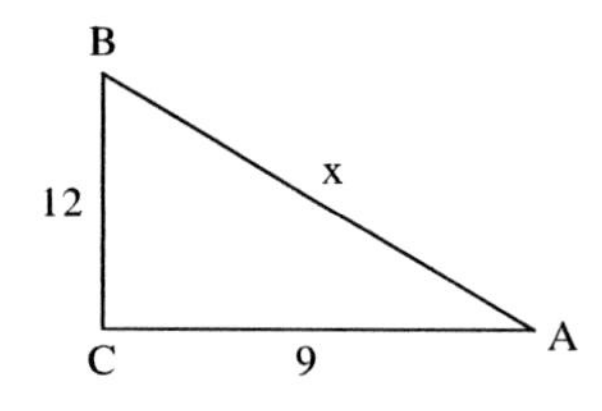

Answer:__________

SECTION 38

TRIGONOMETRIC FUNCTIONS OF AN ACUTE ANGLE

In a right triangle the vertices can be represented as A, B, and C. C is the vertex of the right angle and the sides opposite are a, b, and c. Angle A the side opposite is called "a" and the side adjacent is called "b". Angle B, the side opposite is called "b" and the side adjacent is called "a".

To find the distance c in a right triangle, the Pythagorean theorem can be utilized.

$\text{Leg}^2 + \text{leg}^2 = \text{hyp}^2$

$a^2 + b^2 = c^2$

$\sqrt{c} = \sqrt{a + b}$

$c = \sqrt{a + b}$

The trigonometric function of angle A may be defined using the legs and the hypotenuse of a right triangle.

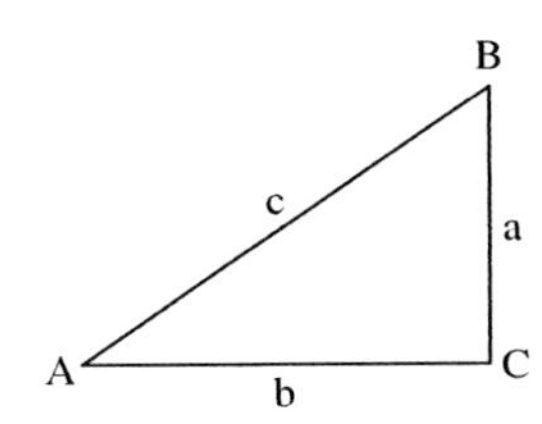

$$\sin A = \frac{\text{opposite side}}{\text{hypotenuse}} = \frac{a}{c}$$

$$\cos A = \frac{\text{adjacent side}}{\text{hypotenuse}} = \frac{b}{c}$$

$$\tan A = \frac{\text{opposite side}}{\text{adjacent side}} = \frac{a}{b}$$

$$\csc A = \frac{\text{hypotenuse}}{\text{opposite side}} = \frac{c}{a}$$

$$\sec A = \frac{\text{hypotenuse}}{\text{adjacent side}} = \frac{c}{b}$$

$$\text{cotan } A = \frac{\text{adjacent side}}{\text{opposite side}} = \frac{b}{a}$$

In the above right triangle, the acute angles are complementary angles (A + B = 90°)

$$\sin B = \frac{b}{c} = \cos A$$

$$\cos B = \frac{a}{c} = \sin A$$

$$\tan B = \frac{b}{a} = \cot A$$

$$\csc B = \frac{c}{b} = \sec A$$

$$\sec B = \frac{c}{a} = \csc A$$

$$\cot B = \frac{a}{b} = \tan A$$

Example: Find the values of the trigonometric functions of the angles of the right triangle

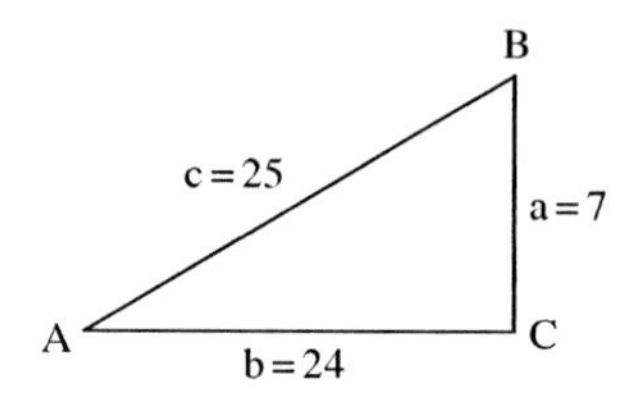

$$\sin A = \frac{\text{opposite side}}{\text{hypotenuse}} = \frac{a}{c} = \frac{7}{25}$$

$$\cos A = \frac{\text{adjacent side}}{\text{hypotenuse}} = \frac{b}{c} = \frac{24}{25}$$

$$\tan A = \frac{\text{opposite side}}{\text{adjacent side}} = \frac{a}{b} = \frac{7}{24}$$

$$\sin B = \frac{\text{opposite side}}{\text{hypotenuse}} = \frac{b}{c} = \frac{24}{25}$$

$$\cos B = \frac{\text{adjacent side}}{\text{hypotenuse}} = \frac{a}{c} = \frac{7}{25}$$

$$\tan B = \frac{\text{opposite side}}{\text{adjacent side}} = \frac{b}{a} = \frac{24}{7}$$

$$\csc A = \frac{\text{hypotenuse}}{\text{opposite side}} = \frac{c}{a} = \frac{25}{7}$$

$$\sec A = \frac{\text{hypotenuse}}{\text{adjacent side}} = \frac{c}{b} = \frac{25}{24}$$

$$\cot A = \frac{\text{adjacent side}}{\text{opposite side}} = \frac{b}{a} = \frac{24}{7}$$

$$\csc B = \frac{\text{hypotenuse}}{\text{opposite side}} = \frac{c}{b} = \frac{25}{24}$$

$$\sec B = \frac{\text{hypotenuse}}{\text{adjacent side}} = \frac{c}{a} = \frac{25}{7}$$

$$\cot B = \frac{\text{adjacent side}}{\text{opposite side}} = \frac{a}{b} = \frac{7}{24}$$

Exercise: For each of the following right triangles, find the values of the trigonometric functions of the angles.

1.

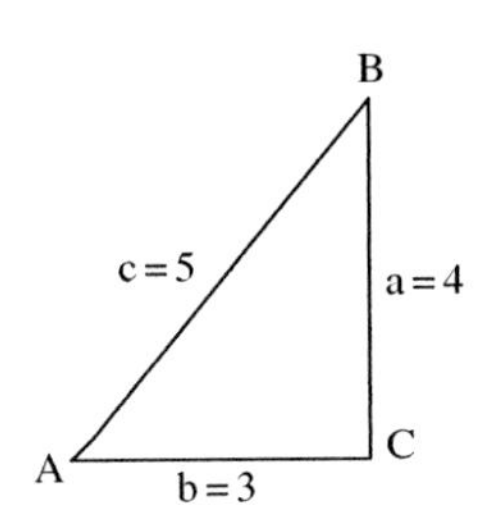

2.

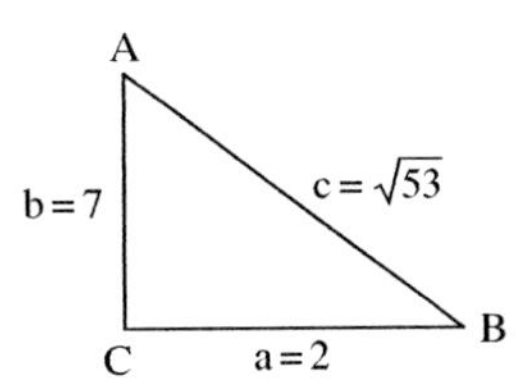

3.

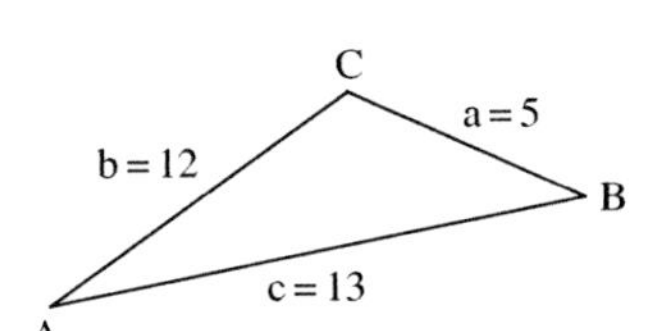

4.

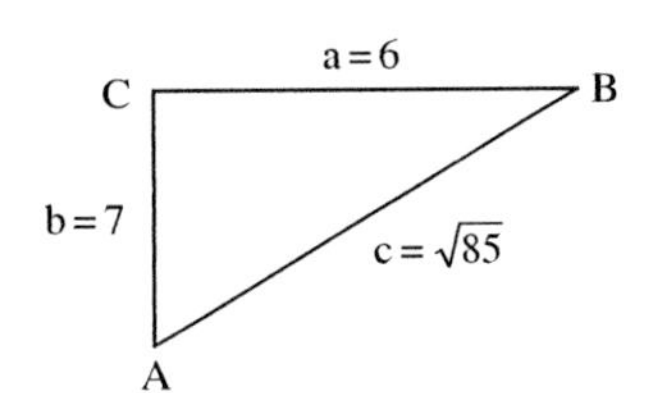

SECTION 39

TRIGONOMETRIC FUNCTIONS OF 30°, 45°, AND 60°

The acute angles 30°, 45°, and 60° have trigonometric functions that can be computed exactly.

Angle θ	$\sin\theta$	$\cos\theta$	$\tan\theta$	$\csc\theta$	$\sec\theta$	$\text{cotan}\,\theta$
30°	$\frac{1}{2}$	$\sqrt{3}$	$\sqrt{3}$	2	$2\sqrt{3}$	$\sqrt{3}$
45°	$\sqrt{2}$	$\sqrt{2}$	1	$\sqrt{2}$	$\sqrt{2}$	1
60°	$\sqrt{3}$	$\frac{1}{2}$	$\frac{\sqrt{3}}{2}$	$2\sqrt{3}$	2	$\sqrt{3}$

Example: Evaluate the following trigonometric functions

1. $\sin 30° + \cos 30°$

   $\frac{1}{2} + \sqrt{3}$ substitute

   $1 + \sqrt{3}$ simplify

2. $\sin 30° \cot 45°$

   $\frac{1}{2} \bullet 1$ substitute

   $\frac{1}{2}$ simplify

Exercise: Evaluate the following trigonometric functions. Fill in the answer in the space provided.

1. $\tan 45° + \sin 30°$ Answer:_________
2. $\cot 45° \tan 45°$ Answer:_________
3. $\cos 30° \sin 60° + \cos 60° \sin 30°$ Answer:_________
4. $\dfrac{\sin 30° \cot 45°}{\sec 60°}$ Answer:_________

5. $\dfrac{\cot 30^\circ + \sec 30^\circ}{\tan 45^\circ}$ Answer:________

6. $\dfrac{\tan 30^\circ + \sec 30^\circ}{\tan 60^\circ}$ Answer:________

7. $\cot 30^\circ \tan 60^\circ$ Answer:________

8. $\csc 30^\circ + \tan 45^\circ$ Answer:________

9. $\dfrac{\tan 45^\circ \cos 60^\circ}{\sin 30^\circ}$ Answer:________

10. $\dfrac{\sin 60^\circ + \sin 30^\circ}{\sec 60^\circ}$ Answer:________

Values of Trigonometric Functions of Acute Angles other than 30°, 45° and 60°

Use the table in the back of the text to determine the values. The table can be used to determine the value of the trigonometric functions as well the value of the angle. Angles less than 45° are located in the left hand column and the functions are read from the top of the page. Angles greater than 45° are located in the right hand column and the functions are read from the bottom of the page.

Exercise: Write the value of the trigonometric function for each of the following. Fill in the correct answer in the space provided.

1. sin 0.7° Answer:________
2. sec 88.2° Answer:________
3. tan 17.5° Answer:________
4. csc 19.7° Answer:________
5. cot 89.7° Answer:________
6. tan 39.8° Answer:________
7. csc 49.2° Answer:________
8. sec 12.7° Answer:________
9. csc 49.8° Answer:________
10. sin 54.10° Answer:________
11. tan 39.7° Answer:________
12. cot 67.8° Answer:________
13. sin 12.2° Answer:________
14. csc 89.2° Answer:________
15. sin 78.6° Answer:________

Exercise: Write the value of the angle that corresponds to the given trigonometric function. Use the table in the back of the text. Fill in the answer in the space provided.

1. $\sin\theta = 0.4067$ Answer:________
2. $\cos\theta = 0.9114$ Answer:________
3. $\sec\theta = 2.1371$ Answer:________
4. $\csc\theta = 1.1145$ Answer:________
5. $\tan\theta = 0.4599$ Answer:________
6. $\text{cotan}\ \theta = 2.0594$ Answer:________
7. $\sec\theta = 2.3751$ Answer:________
8. $\cot\theta = 0.5095$ Answer:________
9. $\cos\theta = 0.8471$ Answer:________
10. $\sin\theta = 0.5548$ Answer:________
11. $\tan\theta = 0.6822$ Answer:________
12. $\csc\theta = 1.7566$ Answer:________
13. $\sin\theta = 0.8121$ Answer:________
14. $\cos\theta = 0.7212$ Answer:________
15. $\tan\theta = 1.4019$ Answer:________

SECTION 41

USING TABLE TO EVALUATE TRIGONOMETRIC FUNCTIONS

The angles are listed in the left and right hand columns. Angles less than 45° are located in the left hand column and angles greater than 45° are located in the right hand column. The table contains angles by degrees and multiples of 10- minute intervals. The table can be used to extrapolate the value of the angle between multiples of 10.

Example :

1. Find $\sin 34° 40'$
   Since $\sin 34° 40'$ is less than 45°, the value is found in the left-hand column.

$$\sin 34° 40' = 0.5688$$

2. Find $\cos 82°$. Since $\cos 82°$ is greater than 45°, the value is found in the right hand column. Read the value in the column labeled cos A at the bottom.

$$\cos 82° = 0.1392$$

3. Find $\sin 14° 43'$

Solution: $\sin 14° 40'' = 0.2532$

$\sin 14° 50' = \underline{0.2560}$

Difference $10' = 0.0028$

Difference for $3' = 0.3\ (0.0028) = 0.00084$

(rounded off to four decimal places) $= 0.0008$

$\sin 14° 43' = 0.2532 + 0.0008 = 0.2540$

4. Find $\cos 74° 26'$

$\cos 74° 20'' = 0.2700$

$\cos 74° 30' = \underline{0.2672}$

Difference $10' = 0.0028$

Difference for $6' = 0.6\ (0.0028) = 0.00168$

(rounded off to four decimal places $= 0.0017$

$\cos 74° 26' = 0.2700 - 0.0017 = 0.2683$

Exercise: Find the function value using thee tables. Fill in the correct answer in the space provided.

1. sin 66° 34′ Answer: ________
2. cos 29° 41 Answer: ________
3. tan 87° 19′ Answer: ________
4. sec 33° 27′ Answer: ________
5. cot 31° 18′ Answer: ________
6. sin 48° 37′ Answer: ________
7. csc71° 3′ Answer: ________
8. tan 72° 18′ Answer: ________
9. cos 18° 29′ Answer: ________
10. csc 21° 18′ Answer: ________
11. sec 39° 21′ Answer: ________
12. sin 18° 18″ Answer: ________

SECTION

42

To solve for the missing angles and sides of a right triangle

Solve the right triangle in which $\angle A = 25° \ 10'$ and $C = 62.5$

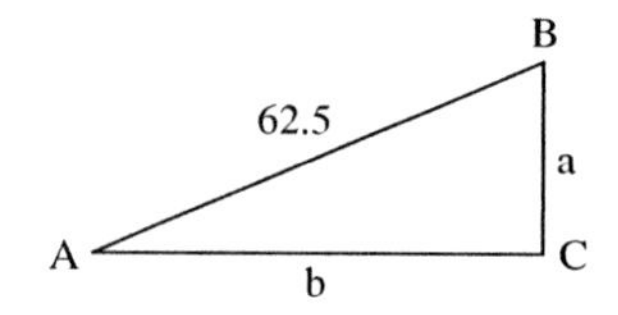

Step I. Find $\angle B$

$\angle B = 90° - 25° \ 10' = 64° \ 50'$

Step II. $\sin \angle A = \dfrac{\text{opposite side}}{\text{hypotenuse}} = \dfrac{a}{c}$

$\sin 25° 10' = \dfrac{a}{62.5}$

$0.4253 = \dfrac{a}{62.5}$

$26.5813 = a$

Step III. $\tan \angle B = \dfrac{b}{a}$

$\tan 64° \ 50' = \dfrac{b}{26.5813}$

$2.1283 = \dfrac{b}{26.5813}$

$56.5730 = \text{b}$

Answer: $\angle B = 64° \ 50'$

a = 26.5813

b = 56.5730

Exercise: Solve the right triangle. Fill in the answers in the space provided.

1.	$\angle A = 31° \, 21'$	$c = 119$	Answer:_________
2.	$\angle B = 58° \, 40'$	$c = 225$	Answer:_________
3.	$\angle A = 33° \, 18'$	$c = 246$	Answer:_________
4.	$\angle B = 54° \, 18'$	$c = 192$	Answer:_________
5.	$\angle A = 68° \, 6'$	$b = 97.2$	Answer:_________
6.	$\angle A = 29° \, 48'$	$b = 458.2$	Answer:_________
7.	$a = 22.3$	$b = 39.7$	Answer:_________
8.	$a = 38.64$	$b = 48.74$	Answer:_________
9.	$a = 506.2$	$c = 984.8$	Answer:_________
10.	$b = 672.9$	$c = 888.1$	Answer:_________
11.	$\angle A = 39° \, 18'$	$b = 107.2$	Answer:_________
12.	$\angle B = 64° \, 18'$	$c = 292$	Answer:_________
13.	$a = 12.79$	$b = 19.84$	Answer:_________
14.	$\angle A = 78° \, 96'$	$b = 197.2$	Answer:_________
15.	$\angle B = 18° \, 18'$	$c = 20.75$	Answer:_________

Problem Set #24

Review Exercises Involving Right Triangles

I. For each of the following right triangles, find the values of the trigonometric functions of the angles:

1.

B
c = 10
a = 8
A
C
b = 6

2.

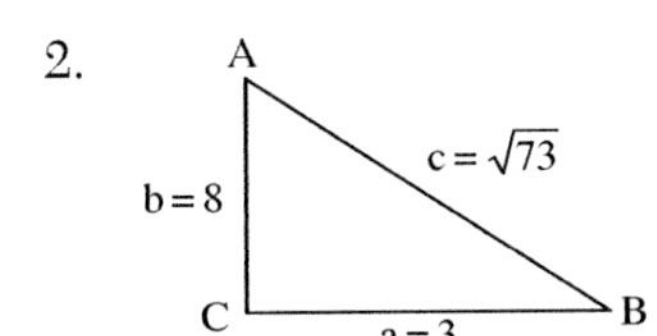

3.

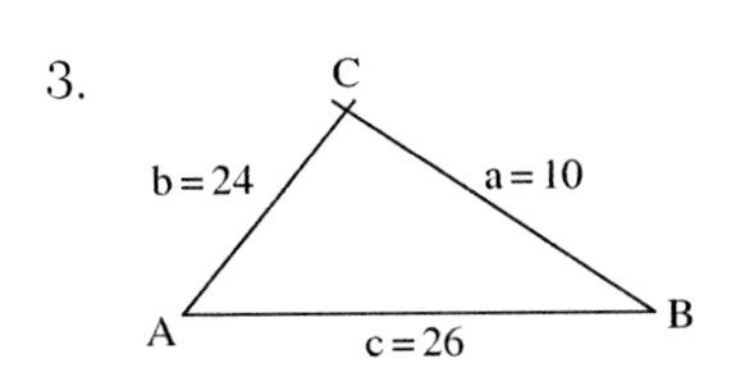

4.

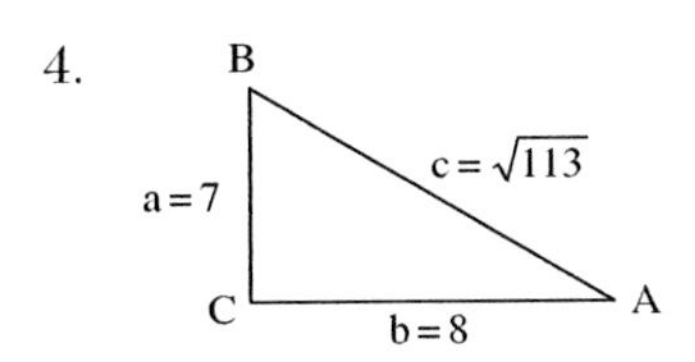

II. Evaluate the following trigonometric functions. Fill in the answer in the space provided.

1. $\tan 30^\circ + \sin 45^\circ$ Answer:________
2. $\sin 60^\circ \cos 60^\circ$ Answer:________
3. $\dfrac{\cos 60^\circ \tan 45^\circ}{\csc 60^\circ}$ Answer:________
4. $\dfrac{\sin 30^\circ + \sin 60^\circ}{\sec 60^\circ}$ Answer:________
5. $\dfrac{\cot 30^\circ + \sec 30^\circ}{\tan 45^\circ}$ Answer:________
6. $\sin 30^\circ + \tan 45^\circ$ Answer:________
7. $\cos 60^\circ \tan 30^\circ$ Answer:________
8. $\cot 60^\circ \tan 30^\circ$ Answer:________
9. $\tan 45^\circ + \sec 45^\circ$ Answer:________
10. $\sin 30^\circ + \sin 60^\circ \sec 60^\circ$ Answer:________

III. Write the value of the trigonometric function for each of the following. Fill in the correct answer in the space provided.

1. $\sin 0.80^\circ$ Answer:________
2. $\sec 89.2^\circ$ Answer:________
3. $\tan 18.5^\circ$ Answer:________
4. $\csc 18.7^\circ$ Answer:________
5. $\cot 79.7^\circ$ Answer:________
6. $\tan 29.8^\circ$ Answer:________
7. $\csc 48.2^\circ$ Answer:________
8. $\sec 13.7^\circ$ Answer:________
9. $\csc 48.7^\circ$ Answer:________
10. $\sin 58.1^\circ$ Answer:________
11. $\tan 29.6^\circ$ Answer:________
12. $\cot 76.8^\circ$ Answer:________
13. $\sin 18.2^\circ$ Answer:________
14. $\csc 79.6^\circ$ Answer:________
15. $\sin 68.7^\circ$ Answer:________

IV. Write the angle that corresponds to the given trigonometric function. Use the table in the back of the text. Fill in the answer in the space provided.

1. $\tan \theta = 1.6447$ Answer:________
2. $\cos \theta = 0.8462$ Answer:________
3. $\sin \theta = 0.4741$ Answer:________
4. $\tan \theta = 1.4071$ Answer:________
5. $\csc \theta = 2.1231$ Answer:________

6. $\sec\theta = 1.7178$ Answer:________
7. $\cot\theta = 0.6873$ Answer:________
8. $\cos\theta = 0.5693$ Answer:________
9. $\sec\theta = 1.7137$ Answer:________
10. $\csc\theta = 1.2329$ Answer:________
11. $\cos\theta = 0.8471$ Answer:________
12. $\sec\theta = 1.1792$ Answer:________
13. $\sin\theta = 0.5505$ Answer:________
14. $\csc\theta = 1.7478$ Answer:________
15. $\tan\theta = 0.6249$ Answer:________

V. Find the function value using the table. Fill in the correct answer in the space provided.

1. $\sin 16^\circ\ 16'$ Answer:________
2. $\cos 39^\circ\ 41'$ Answer:________
3. $\tan 78^\circ\ 17'$ Answer:________
4. $\sec 43^\circ\ 27'$ Answer:________
5. $\cot 18^\circ\ 18'$ Answer:________
6. $\sin 16^\circ\ 16'$ Answer:________
7. $\csc 61^\circ\ 4'$ Answer:________
8. $\tan 82^\circ\ 18'$ Answer:________
9. $\cos 28^\circ\ 39'$ Answer:________
10. $\cot 39^\circ\ 17'$ Answer:________
11. $\sin 67^\circ\ 19'$ Answer:________
12. $\cot 18^\circ\ 18'$ Answer:________

VI. Solve the right triangle. Fill in the answer in the space provided.

1. $\angle A = 71^\circ\ 20'$ $c = 105$ Answer:________
2. $\angle B = 48^\circ\ 40'$ $c = 175$ Answer:________
3. $a = 12.3$ $b = 29.7$ Answer:________
4. $b = 17$ $c = 29$ Answer:________
5. $\angle A = 74^\circ\ 18'$ $b = 187.2$ Answer:________
6. $a = 32.3$ $b = 49.7$ Answer:________
7. $\angle B = 18^\circ\ 18'$ $c = 30.75$ Answer:________
8. $\angle A = 68^\circ\ 16'$ $b = 107.2$ Answer:________
9. $a = 48.74$ $a = 38.64$ Answer:________
10. $a = 39.7$ $b = 32.3$ Answer:________

Reducing Trigonometric Functions To Positive Acute Angles

SECTION

43

Let θ represent any angle. The following is known:

$$\sin(\theta + n\,360°) = \sin\theta \qquad \csc(\theta + n\,360°) = \csc\theta$$
$$\cos(\theta + n\,360°) = \cos\theta \qquad \sec(\theta + n\,360°) = \sec\theta$$
$$\tan(\theta + n\,360°) = \tan\theta \qquad \cot(\theta + n\,360°) = \cot\theta$$

n is any positive or negative integer or zero.

Examples: Express each of the following as a function of a reduced angle.

1. $\cos 400° = (\cos(40° + 36°) = \cos\ 40°$
2. $\cos 860° = (\cos(140 + 2 \bullet 360°) = 140°$
3. $\tan(-1010°) = (\tan(70°\ 3 \bullet 360°) = \tan 70°$

Examples: Express each of the following as a function of an acute angle.

1. $\cos 322° = \cos(360° - 322°) = \cos 38°$
2. $\tan 902° = \tan(182° + 2(360°)) = \tan 180°$

   $\tan(182° - 180°) = \tan 2°$
3. $\sin 233° = \sin(233° - 180°) = -\sin 52°$

Exercise: Express each of the following as a function of an acute angle. Fill in the answer in the space provided.

1. $\sin 244°$ Answer:________
2. $\tan 910°$ Answer:________
3. $\cos 324°$ Answer:________
4. $\sin 230°$ Answer:________
5. $\tan 325°$ Answer:________
6. $\csc 875°$ Answer:________
7. $\tan 165°$ Answer:________

8. sec 260° Answer:_________
9. sin 670° Answer:_________
10. cot 930° Answer:_________
11. sin 320° Answer:_________
12. tan 235° Answer:_________
13. csc 785° Answer:_________
14. tan 175° Answer:_________
15. sec 270° Answer:_________

SECTION 44

FUNCTIONS OF A NEGATIVE ANGLE

$\sin(-\theta) = -\sin\theta$

$\cos(-\theta) = \cos\theta$

$\tan(-\theta) = -\tan\theta$

$\cot(-\theta) = -\cot\theta$

$\sec(-\theta) = \sec\theta$

$\csc(-\theta) = -\csc\theta$

Examples: Rewrite the following functions of a negative angle.

1. $\sin(-60°) = -\sin 60°$
2. $\tan(-210°) = -\tan 210°$
3. $\cos(-40°) = \cos 40°$

Exercise: Rewrite the following functions of a negative angle. Fill in the answer in the space provided.

1. $\sec(-70°)$ Answer: ________
2. $\cos(-220°)$ Answer: ________
3. $\sin(-230°)$ Answer: ________
4. $\cos(-70°)$ Answer: ________
5. $\tan(-50°)$ Answer: ________
6. $\cot(-210°)$ Answer: ________
7. $\csc(-320°)$ Answer: ________
8. $\sec(-60°)$ Answer: ________
9. $\sin(-70°)$ Answer: ________
10. $\cos(-80°)$ Answer: ________
11. $\tan(-230°)$ Answer: ________
12. $\cot(-190°)$ Answer: ________

13. sec $(-160°)$ Answer: __________
14. csc $(-260°)$ Answer: __________
15. tan $(-290°)$ Answer: __________

Reference Angles

SECTION 45

Reference angles can be found for an angle from 0° to 360°. A reference angle R for angle θ can be represented as an acute angle. The following table can be used to find the value of trigonometric functions of any angle.

Quadrants

II	I
III	IV

Quadrant for θ	Relationship	Function Signs
I	$R = \theta$	All functions are positive
II	$R = 180° - \theta$	Only sin R and csc R are positive
III	$R = \theta - 180°$	Only tan R and cot R are positive
IV	$R = 360° - \theta$	Only cos R and sec R are positive

Express each of the following as a function of a reduced angle; fill in the correct answer in the space provided.

1. sin 400° Answer:________
2. cos 840° Answer:________
3. tan (−1000)° Answer:________
4. sin 390° Answer:________
5. cos 760° Answer:________
6. sec 420° Answer:________
7. csc 390° Answer:________

8. $\tan(-1380)^\circ$ Answer:_________

9. $\csc 390^\circ$ Answer:_________

10. $\cos 430^\circ$ Answer:_________

11. $\sin 420^\circ$ Answer:_________

12. $\cos 850^\circ$ Answer:_________

13. $\cos 770^\circ$ Answer:_________

14. $\sin 390^\circ$ Answer:_________

15. $\tan 255^\circ$ Answer:_________

PROBLEM SET #25

Review Exercise involving Functions of Angles:

1. $\sin 390^\circ$ Answer:_________

2. $\tan(-1005)^\circ$ Answer:_________

3. $\cos 850^\circ$ Answer:_________

4. $\sin 375^\circ$ Answer:_________

5. $\sec 430^\circ$ Answer:_________

6. $\tan(-1370)^\circ$ Answer:_________

7. $\cot 420^\circ$ Answer:_________

8. $\cos 830^\circ$ Answer:_________

9. $\sin 375^\circ$ Answer:_________

10. $\cos 780^\circ$ Answer:_________

11. $\cot 390^\circ$ Answer:_________

12. $\sin 320^\circ$ Answer:_________

II. Rewrite the following functions of a negative angle: Fill in the answer in the space provided.

1. $\sec(-80^\circ)$ Answer:_________

2. $\cos(-230^\circ)$ Answer:_________

3. $\sin(-240^\circ)$ Answer:_________

4. $\tan(-60^\circ)$ Answer:_________

5. $\cot(-205^\circ)$ Answer:_________

6. $\csc(-130^\circ)$ Answer:_________

7. $\sec(-70^\circ)$ Answer:_________

8. $\tan(-240^\circ)$ Answer:________

9. $\sec(-170^\circ)$ Answer:________

10. $\csc(-265^\circ)$ Answer:________

11. $\tan(-230^\circ)$ Answer:________

12. $\sec(-70^\circ)$ Answer:________

13. $\csc(-130^\circ)$ Answer:________

14. $\sin(-70^\circ)$ Answer:________

15. $\csc(-290^\circ)$ Answer:________

1. $\sin 254^\circ$ Answer:________

2. $\tan 932^\circ$ Answer:________

3. $\cos 234^\circ$ Answer:________

4. $\sin 234^\circ$ Answer:________

5. $\tan 235^\circ$ Answer:________

6. $\sec 785^\circ$ Answer:________

7. $\tan 265^\circ$ Answer:________

8. $\sec 160^\circ$ Answer:________

9. $\sin 760^\circ$ Answer:________

10. $\cot 390^\circ$ Answer:________

11. $\sin 230^\circ$ Answer:________

12. $\tan 325^\circ$ Answer:________

13. $\csc 875^\circ$ Answer:________

14. $\tan 185^\circ$ Answer:________

15. $\sec 260^\circ$ Answer:________

SECTION 46

Basic Relationship and Identities

Basic Relationships

Reciprocal Relationship

$$\csc\theta = \frac{1}{\sin\theta}$$

$$\sec\theta = \frac{1}{\cos\theta}$$

$$\cot\theta = \frac{1}{\cot\theta}$$

Quotient Relationship

$$\tan\theta = \frac{\sin\theta}{\cos\theta}$$

$$\cot\theta = \frac{\cos\theta}{\sin\theta}$$

Pythagorean Relationship:

$$\sin^2\theta + \cos^2\theta = 1$$

$$1 + \tan^2\theta = \sec^2\theta$$

$$1 + \cot^2\theta = \csc^2\theta$$

Solving Trigonometric Identities

What are Trigonometric Identities? They are equations involving trigonometric functions which hold true for all values of the angles which the functions are defined.

Suggested Guidelines for Proving Identities

To prove a trigonometric identity, transform one of the members into another. Begin with the more complicated side. Transform each side into the same form.

1. Remember the eight basic trigonometric relationships and the different forms of each relationship.
2. Understand and be knowledgeable of various forms of factoring as well as different techniques to find the product.
3. Have a strong conceptual knowledge of all the processes involving functions and transforming functions into equivalent ones.
4. Use the process of substitution on each side of the identity.
5. Try to avoid using radicals.
6. Treat each side of the identity separately trying to put both sides into the same form.
7. Work with the more complicated side of the identity trying to transform it into the form of the other side.
8. Try to convert all trigonometric functions into functions of only sine and cosine.
9. It may help to simplify the identity by multiplying the numerator and the denominator of the fraction by the conjugate.
10. Use conjugates to simplify the square root of a function to transform it into perfect squares.

Example: Verify the identity

$$\cos\theta \csc\theta = \cot\theta$$

Solution: Transform both sides of the identity into sine and cosine. Use these two identities

$$\csc\theta = \frac{1}{\sin\theta}$$

$$\cot\theta = \frac{\cos\theta}{\sin\theta}$$

and substitute their values into the identity to be verified.

$$\cos\theta \csc\theta = \cot\theta$$

$$\cos\theta \bullet \frac{1}{\sin\theta} = \frac{\cos\theta}{\sin\theta}$$

$$\frac{\cos\theta}{\sin\theta} = \cot\theta$$

Example: Verify the identity

$$\frac{\sin\theta}{\csc\theta} + \frac{\cos\theta}{\sec\theta} = 1$$

Solution: Transform both sides of the identity into sine and cosine. Use the following two reciprocal relationships.

$$\csc\theta = \frac{1}{\sin\theta}$$

$$\sec\theta = \frac{1}{\cos\theta}$$

and substitute their values into the identity to be verified.

$$\frac{\sin\theta}{\csc\theta} + \frac{\cos\theta}{\sec\theta} = 1$$

$$\frac{\sin\theta}{\frac{1}{\sin\theta}} + \frac{\cos\theta}{\frac{1}{\cos\theta}} = 1$$ Substitution

$$\sin^2\theta + \cos^2\theta = 1$$ Divide and find quotient

Now use Pythagorean Relationship

$$\sin^2\theta + \cos^2\theta = 1$$

$$1 = 1$$ Substitute Pythagorean relationship

Exercise: Verify the following identities: Show all work in the space provided.

1. $\cos\theta \csc\theta = \cot\theta$
2. $\dfrac{\sin\theta}{\csc\theta} + \dfrac{\cos\theta}{\sec\theta} = 1$
3. $\cot^2 A \sin^2 A \tan^2 \csc^2 A = 1$
4. $\dfrac{(\sin^2 X + \cos^2 X)\cos^2 X}{\sin^2 X} = \cot^2 X$
5. $(1 - \cos^2 A)\csc^2 A = 1$
6. $\dfrac{\csc\theta + \sec\theta}{\cot\theta + \tan\theta} = \cos\theta + \sin\theta$
7. $\sin X \tan X + \cos X = \sec X$
8. $\dfrac{1}{\tan X \sec X} = \csc X - \sin X$
9. $\dfrac{1}{\cot\theta} + \dfrac{1}{\tan\theta} = \cot\theta + \tan\theta$
10. $(1 - \sin\theta)(1 + \sin\theta) = \cos^2\theta$
11. $\csc X - \sin X = \dfrac{\cot^2 X}{\csc X}$
12. $\tan\theta \sin\theta + \cos\theta = \sec\theta$
13. $\cos X \sec X = 1$
14. $\dfrac{1}{\tan X \sec X} = \csc X - \sin X$
15. $\dfrac{\tan X + \sin X}{\csc X + \cot X} = \tan X \sin X$

Trigonometric Functions Sum – Difference – Product

SECTION 47

I. Product of Sines and Cosines: $\alpha\beta$

$$\sin\alpha\cos\beta = \frac{1}{2}[\sin(\alpha+\beta) + \sin(\alpha-\beta)]$$

$$\cos\alpha\sin\beta = \frac{1}{2}[\sin(\alpha+\beta) - \sin(\alpha-\beta)]$$

$$\cos\alpha\cos\beta = \frac{1}{2}[\cos(\alpha+\beta) + \cos(\alpha-\beta)]$$

$$\sin\alpha\sin\beta = -\frac{1}{2}[\cos(\alpha+\beta) - \cos(\alpha-\beta)]$$

II. Sum and Differences of Sines and Cosines:

$$\sin A + \sin B = 2\sin\frac{1}{2}(A+B)\cos\frac{1}{2}(A-B)$$

$$\sin A - \sin B = 2\cos\frac{1}{2}(A+B)\sin\frac{1}{2}(A-B)$$

$$\cos A + \cos B = 2\cos\frac{1}{2}(A+B)\cos\frac{1}{2}(A-B)$$

$$\cos A - \cos B = -2\sin\frac{1}{2}(A+B)\sin\frac{1}{2}(A-B)$$

Example: Express each of the following as a sum or difference of sines or cosines:

1. $\sin 40^\circ \cos 30^\circ$

$$\sin 40^\circ\cos 30^\circ = \frac{1}{2}[\sin(40^\circ+30^\circ) + \sin(40^\circ-30^\circ)]$$
$$= \frac{1}{2}(\sin 70^\circ + \sin 10^\circ)$$

2. $\cos 60^\circ \cos 40^\circ$

$$\cos 60^\circ\cos 40^\circ = \frac{1}{2}[\cos(60^\circ+40^\circ) + \cos(60^\circ-40^\circ)]$$
$$= \frac{1}{2}(\cos 100^\circ + \cos 20^\circ)$$

Examples: Express each of the following as a product of sines and cosines:

1. $\sin 40° + \sin 30° = 2 \sin \frac{1}{2}(40° + 30°) \cos \frac{1}{2}(40° - 30°)$
$= 2 \sin 35° \cos 5°$

2. $\cos 50° + \cos 20° = 2 \cos \frac{1}{2}(50° + 20°) \cos \frac{1}{2}(50° - 20°)$
$= 2 \cos 35° \cos 15°$

Exercise: Express each of the following as a sum or difference of sines and cosines: Fill in the answer in the space provided.

1. $\sin 35° \cos 40°$ Answer:________
2. $\sin 2X \cos 4X$ Answer:________
3. $\sin 60° \sin 30°$ Answer:________
4. $\sin 50° \cos 20°$ Answer:________
5. $\sin 5X \sin 4X$ Answer:________
6. $\cos 30° \sin 50°$ Answer:________
7. $\sin 140° \sin 60°$ Answer:________
8. $\sin 4X \cos 2X$ Answer:________
9. $\sin 20° \cos 30°$ Answer:________
10. $\sin 30° \cos 40°$ Answer:________

Exercise: Express each of the following as a product of sines and cosines: Fill in the answer in the space provided.

1. $\sin 40° + \sin 20°$ Answer:________
2. $\sin 75° - \sin 35°$ Answer:________
3. $\cos 55° + \cos 25°$ Answer:________
4. $\cos 60° + \cos 20°$ Answer:________
5. $\cos 45° + \cos 15°$ Answer:________
6. $\sin 85° - \sin 45°$ Answer:________
7. $\sin 70° - \sin 30°$ Answer:________
8. $\cos 30° + \cos 10°$ Answer:________
9. $\sin 20° + \sin 10°$ Answer:________
10. $\sin 50° + \sin 30°$ Answer:________

SECTION 48

Solving Triangles That Do Not Have a Right Angle — Oblique Triangles

In this section we will continue our study of triangles and their usefulness as mathematical models in many situations. Specifically we will consider triangles that are not right triangles. The convention of denoting the angles by A, B, C and the lengths of the corresponding opposite sides by a, b, c will be used here.

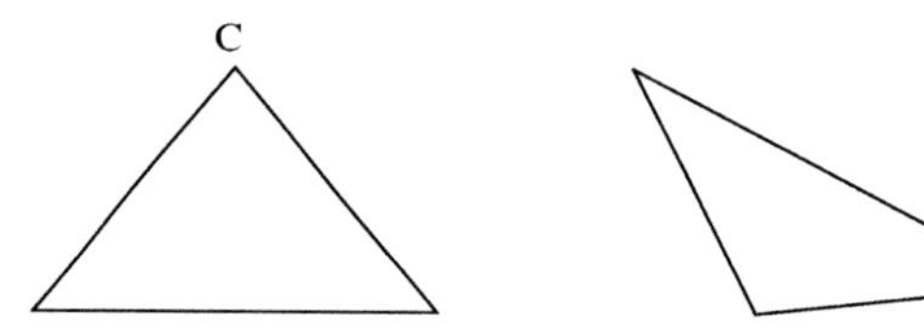

When three parts, not all angles are known, the triangle is uniquely determined, except in one case to be noted below, the four cases of these types of triangles are:

Case 1. Given one side and two angles.

Case 2. Given two sides and the angle opposite one of them.

Case 3. Given two sides and the included angle.

Case 4. Given the three sides.

The LAW OF SINES:In any triangle, the sides are proportional to the sines of the opposite angles. For example,

$$\frac{\sin A}{a} = \frac{\sin B}{b} = \frac{\sin C}{c}$$

In using the Law of Sines, all you need to do is to set any two of these relationship equal to each other. For example,

$$\frac{\sin A}{a} = \frac{\sin B}{b} \quad \text{or} \quad \frac{\sin B}{b} = \frac{\sin C}{c} \quad \text{or} \quad \frac{\sin A}{a} = \frac{\sin C}{c}$$

Example 1: In triangle ABC, $A = 30°$, $B = 70°$, and $a = 8$ ft. Find side c.

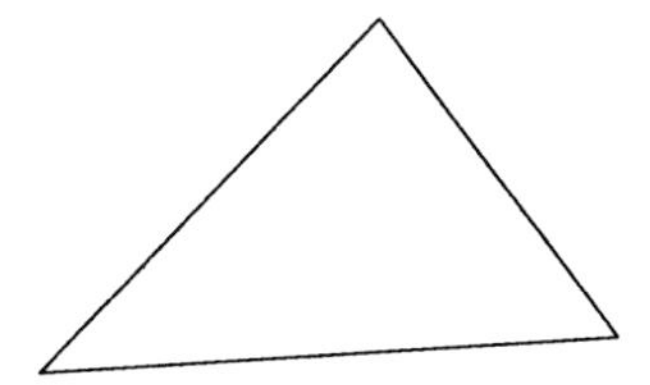

We can see that angle C = 80°, since all three angles must add up to 180°, we can say that $A + B + C = 180°$

$$30° + 70° + C = 180°; \text{ therefore}$$

$$100° + C = 180°$$

$$C = 80°$$

We can then say that

$$\frac{\sin C}{c} = \frac{\sin A}{a}$$

To solve for c, we multiply both sides in sinC and then substitute.

$$C = \frac{a \sin C}{\sin A}$$

$$C = \frac{8 \sin 80°}{\sin 30°}$$

$$C = \frac{8(0.9848)}{0.5000}$$

$$C = 15.7568 \text{ ft.}$$

Example 2: Solve the triangle ABC if B =82°, C = 34° and a = 6 ft.
Find angle A

$$A = 180° - (B + C)$$

$$A = 180° - (82° + 34°)$$

$$A = 64$$

Find side b, using the following relationship

$$\frac{\sin B}{b} = \frac{\sin A}{a}$$

$$\frac{\sin 82°}{b} = \frac{\sin 64°}{6}$$

You can then cross multiply $b \sin 64° = 6 \sin 82°$
Divide both sides by sin 64°
Then you will get

$$b = \frac{6 \sin 82°}{\sin 64°}$$

then find sin 82° and sin 64°

$$b = \frac{6(0.9903)}{0.8988}$$

$$b = 6.61$$

Find side c

Use the relationship

$$\frac{\sin A}{a} = \frac{\sin C}{c}$$

$\frac{\sin 64°}{6} = \frac{\sin 34°}{c}$, then cross multiply

$c \sin 64° = 6 \sin 34°$, then divide both sides by sin 64°.

$$c = \frac{6 \sin 34°}{\sin 64°}$$

now look up sin 34° and sin 64°.

$$c = \frac{6(0.5592)}{0.8988}$$

$$c = 3.73$$

Problem Set

Use Triangle ABC and find the missing angle or side.

1. $A = 50°$, $B = 70°$, $a = 10$ inches, find b. Answer:________
2. $A = 85°$, $B = 25°$, $b = 8$ inches, find a. Answer:________
3. $A = 38°$, $B = 70°$, $b = 20$ inches, find c. Answer:________
4. $B = 59°$, $C = 65°$, $a = 8$ inches, find b. Answer:________
5. $A = 15°$, $B = 108°$, $a = 14$ inches, find c. Answer:________
6. $A = 60°$, $B = 80°$, $a = 20$ inches, find b. Answer:________
7. $A = 75°$, $B = 35°$, $b = 10$ inches, find a. Answer:________
8. $A = 48°$, $B = 50°$, $b = 20$ inches, find a. Answer:________
9. $B = 69°$, $A = 47°$, $a = 10$ inches, find b. Answer:________
10. $B = 39°$, $A = 62°$, $b = 20$ inches, find a. Answer:________
11. $B = 25°$, $C = 98°$, $a = 18$ inches, find c. Answer:________
12. $B = 69°$, $C = 55°$, $a = 9$ inches, find b. Answer:________
13. $A = 48°$, $C = 60°$, $b = 20$ inches, find c. Answer:________
14. $A = 40°$, $B = 80°$, $a = 10$ inches, find b. Answer:________
15. $A = 48°$, $C = 80°$, $b = 20$ inches, find c. Answer:________

Law of Cosines

We will now look at another relationship that exists between the sides and angles in any triangle. It is called the Law of Cosines and is stated like this.

$$a^2 = b^2 + c^2 - 2bc \cos A$$
$$b^2 = a^2 + c^2 - 2ac \cos B$$
$$c^2 = a^2 + b^2 - 2ab \cos C$$

We can also state the Law of Cosines like this:

$$\cos A = \frac{b^2 + c^2 - a^2}{2bc}$$
$$\cos B = \frac{a^2 + c^2 - b^2}{2ac}$$
$$\cos C = \frac{a^2 + b^2 - c^2}{2ab}$$

Example 1: Given triangle ABC with A = 60°, b = 20 inches, c = 30 inches, find side a.

We can use the relationship

$$a^2 = b^2 + c^2 - 2bc \cos A$$
$$a^2 = 20^2 + 30^2 - 2(20)(30) \cos 60°$$
$$a^2 = 400 + 900 - 1200(0.5)$$
$$a^2 = 700$$
$$a = 26 \text{ inches} \qquad \text{(rounded off to the nearest inch.)}$$

Example 2: Given triangle ABC with a = 5 inches, b = 7 inches, c = 9 inches, find angle A.

We will use the relationship

$$\cos A = \frac{b^2 + c^2 - a^2}{2bc}$$

We then will get

$$\cos A = \frac{7^2 + 9^2 - 5^2}{2(7)(9)}$$
$$\cos A = \frac{49 + 81 - 25}{126}$$
$$\cos A = 0.8333$$
$$\text{Angle } A = 56° \qquad \text{(to the nearest degree)}$$

Problem Set

Solve the following triangles ABC given the following information.

1. a = 125, A = 54°, B = 65°.
2. b = 60, A = 75°, C = 40°.
3. a = 20, b = 35, B = 50°.

4. $a = 24$, $b = 19$, $c = 26$.
5. $a = 6$, $b = 7$, $C = 10°$.
6. If $a = 90$, $b = 50$, $C = 60°$, find c.
7. If $a = 18$, $b = 20$, $c = 22$, find the largest angle.
8. If $b = 4$, $c = 7$, $A = 112°$, find a.
9. If $a = 15$, $b = 9$, $c = 12$, find the smallest angle.
10. If $a = 4$, $b = 3$, $c = 5$, find angle A.

Table 1 Trigonometric Functions—Angle in 10-Minute Intervals

A	sin A	cos A	tan A	cot A	sec A	csc A	
0° 0′	0.0000	1.0000	0.0000	Undefined	1.0000	Undefined	90° 0′
0°10′	0.0029	1.0000	0.0029	343.7730	1.0000	343.7740	89°50′
0°20′	0.0058	1.0000	0.0058	171.8850	1.0000	171.8880	89°40′
0°30′	0.0087	1.0000	0.0087	114.5880	1.0000	114.5930	89°30′
0°40′	0.0116	0.9999	0.0116	85.9396	1.0001	85.9454	89°20′
0°50′	0.0145	0.9999	0.0145	68.7499	1.0001	68.7572	89°10′
1° 0′	0.0175	0.9998	0.0175	57.2901	1.0002	57.2989	89° 0′
1°10′	0.0204	0.9998	0.0204	49.1040	1.0002	49.1142	88°50′
1°20′	0.0233	0.9997	0.0233	42.9641	1.0003	42.9758	88°40′
1°30′	0.0262	0.9997	0.0262	38.1885	1.0003	38.2016	88°30′
1°40′	0.0291	0.9996	0.0291	34.3678	1.0004	34.3823	88°20′
1°50′	0.0320	0.9995	0.0320	31.2416	1.0005	31.2576	88°10′
2° 0′	0.0349	0.9994	0.0349	28.6363	1.0006	28.6537	88° 0′
2°10′	0.0378	0.9993	0.0378	26.4316	1.0007	26.4505	87°50′
2°20′	0.0407	0.9992	0.0407	24.5418	1.0008	24.5621	87°40′
2°30′	0.0436	0.9990	0.0437	22.9038	1.0010	22.9256	87°30′
2°40′	0.0465	0.9989	0.0466	21.4704	1.0011	21.4937	87°20′
2°50′	0.0494	0.9988	0.0495	20.2055	1.0012	20.2303	87°10′
3° 0′	0.0523	0.9986	0.0524	19.0811	1.0014	19.1073	87° 0′
3°10′	0.0552	0.9985	0.0553	18.0750	1.0015	18.1026	86°50′
3°20′	0.0581	0.9983	0.0582	17.1694	1.0017	17.1984	86°40′
3°30′	0.0610	0.9981	0.0612	16.3499	1.0019	16.3804	86°30′
3°40′	0.0640	0.9980	0.0641	15.6048	1.0021	15.6368	86°20′
3°50′	0.0669	0.9978	0.0670	14.9244	1.0022	14.9579	86°10′
	cos A	sin A	cot A	tan A	csc A	sec A	A

A	sin A	cos A	tan A	cot A	sec A	csc A	
4° 0′	0.0698	0.9976	0.0699	14.3007	1.0024	14.3356	86° 0′
4°10′	0.0727	0.9974	0.0729	13.7267	1.0027	13.7631	85°50′
4°20′	0.0756	0.9971	0.0758	13.1969	1.0029	13.2347	85°40′
4°30′	0.0785	0.9969	0.0787	12.7062	1.0031	12.7455	85°30′
4°40′	0.0814	0.9967	0.0816	12.2505	1.0033	12.2912	85°20′
4°50′	0.0843	0.9964	0.0846	11.8262	1.0036	11.8684	85°10′
5° 0′	0.0872	0.9962	0.0875	11.4301	1.0038	11.4737	85° 0′
5°10′	0.0901	0.9959	0.0904	11.0594	1.0041	11.1046	84°50′
5°20′	0.0929	0.9957	0.0934	10.7119	1.0043	10.7585	84°40′
5°30′	0.0958	0.9954	0.0963	10.3854	1.0046	10.4334	84°30′
5°40′	0.0987	0.9951	0.0992	10.0780	1.0049	10.1275	84°20′
5°50′	0.1016	0.9948	0.1022	9.7882	1.0052	9.8391	84°10′
6° 0′	0.1045	0.9945	0.1051	9.5144	1.0055	9.5668	84° 0′
6°10′	0.1074	0.9942	0.1080	9.2553	1.0058	9.3092	83°50′
6°20′	0.1103	0.9939	0.1110	9.0098	1.0061	9.0652	83°40′
6°30′	0.1132	0.9936	0.1139	8.7769	1.0065	8.8337	83°30′
6°40′	0.1161	0.9932	0.1169	8.5555	1.0068	8.6138	83°20′
6°50′	0.1190	0.9929	0.1198	8.3450	1.0072	8.4047	83°10′
7° 0′	0.1219	0.9925	0.1228	8.1444	1.0075	8.2055	83° 0′
7°10′	0.1248	0.9922	0.1257	7.9530	1.0079	8.0156	82°50′
7°20′	0.1276	0.9918	0.1287	7.7704	1.0082	7.8344	82°40′
7°30′	0.1305	0.9914	0.1317	7.5958	1.0086	7.6613	82°30′
7°40′	0.1334	0.9911	0.1346	7.4287	1.0090	7.4957	82°20′
7°50′	0.1363	0.9907	0.1376	7.2687	1.0094	7.3372	82°10′
8° 0′	0.1392	0.9903	0.1405	7.1154	1.0098	7.1853	82° 0′
8°10′	0.1421	0.9899	0.1435	6.9682	1.0102	7.0396	81°50′
8°20′	0.1449	0.9894	0.1465	6.8269	1.0107	6.8998	81°40′
8°30′	0.1478	0.9890	0.1495	6.6912	1.0111	6.7655	81°30′
8°40′	0.1507	0.9886	0.1524	6.5606	1.0116	6.6363	81°20′
8°50′	0.1536	0.9881	0.1554	6.4348	1.0120	6.5121	81°10′
9° 0′	0.1564	0.9877	0.1584	6.3138	1.0125	6.3925	81° 0′
9°10′	0.1593	0.9872	0.1614	6.1970	1.0129	6.2772	80°50′
9°20′	0.1622	0.9868	0.1644	6.0844	1.0134	6.1661	80°40′
9°30′	0.1650	0.9863	0.1673	5.9758	1.0139	6.0589	80°30′
9°40′	0.1679	0.9858	0.1703	5.8708	1.0144	5.9554	80°20′
9°50′	0.1708	0.9853	0.1733	5.7694	1.0149	5.8554	80°10′
10° 0′	0.1736	0.9848	0.1763	5.6713	1.0154	5.7588	80° 0′
10°10′	0.1765	0.9843	0.1793	5.5764	1.0160	5.6653	79°50′
10°20′	0.1794	0.9838	0.1823	5.4845	1.0165	5.5749	79°40′
10°30′	0.1822	0.9833	0.1853	5.3955	1.0170	5.4874	79°30′
10°40′	0.1851	0.9827	0.1883	5.3093	1.0176	5.4026	79°20′
10°50′	0.1880	0.9822	0.1914	5.2257	1.0181	5.3205	79°10′
	cos A	sin A	cot A	tan A	csc A	sec A	A

A	sin A	cos A	tan A	cot A	sec A	csc A	
11° 0′	0.1908	0.9816	0.1944	5.1446	1.0187	5.2408	79° 0′
11°10′	0.1937	0.9811	0.1974	5.0658	1.0193	5.1636	78°50′
11°20′	0.1965	0.9805	0.2004	4.9894	1.0199	5.0886	78°40′
11°30′	0.1994	0.9799	0.2035	4.9152	1.0205	5.0159	78°30′
11°40′	0.2022	0.9793	0.2065	4.8430	1.0211	4.9452	78°20′
11°50′	0.2051	0.9787	0.2095	4.7729	1.0217	4.8765	78°10′
12° 0′	0.2079	0.9781	0.2126	4.7046	1.0223	4.8097	78° 0′
12°10′	0.2108	0.9775	0.2156	4.6382	1.0230	4.7448	77°50′
12°20′	0.2136	0.9769	0.2186	4.5736	1.0236	4.6817	77°40′
12°30′	0.2164	0.9763	0.2217	4.5107	1.0243	4.6202	77°30′
12°40′	0.2193	0.9757	0.2247	4.4494	1.0249	4.5604	77°20′
12°50′	0.2221	0.9750	0.2278	4.3897	1.0256	4.5022	77°10′
13° 0′	0.2250	0.9744	0.2309	4.3315	1.0263	4.4454	77° 0′
13°10′	0.2278	0.9737	0.2339	4.2747	1.0270	4.3901	76°50′
13°20′	0.2306	0.9730	0.2370	4.2193	1.0277	4.3362	76°40′
13°30′	0.2334	0.9724	0.2401	4.1653	1.0284	4.2837	76°30′
13°40′	0.2363	0.9717	0.2432	4.1126	1.0291	4.2324	76°20′
13°50′	0.2391	0.9710	0.2462	4.0611	1.0299	4.1824	76°10′
14° 0′	0.2419	0.9703	0.2493	4.0108	1.0306	4.1336	76° 0′
14°10′	0.2447	0.9696	0.2524	3.9617	1.0314	4.0859	75°50′
14°20′	0.2476	0.9689	0.2555	3.9136	1.0321	4.0394	75°40′
14°30′	0.2504	0.9681	0.2586	3.8667	1.0329	3.9939	75°30′
14°40′	0.2532	0.9674	0.2617	3.8208	1.0337	3.9495	75°20′
14°50′	0.2560	0.9667	0.2648	3.7760	1.0345	3.9061	75°10′
15° 0′	0.2588	0.9659	0.2679	3.7321	1.0353	3.8637	75° 0′
15°10′	0.2616	0.9652	0.2711	3.6891	1.0361	3.8222	74°50′
15°20′	0.2644	0.9644	0.2742	3.6470	1.0369	3.7817	74°40′
15°30′	0.2672	0.9636	0.2773	3.6059	1.0377	3.7420	74°30′
15°40′	0.2700	0.9628	0.2805	3.5656	1.0386	3.7032	74°20′
15°50′	0.2728	0.9621	0.2836	3.5261	1.0394	3.6652	74°10′
16° 0′	0.2756	0.9613	0.2867	3.4874	1.0403	3.6280	74° 0′
16°10′	0.2784	0.9605	0.2899	3.4495	1.0412	3.5915	73°50′
16°20′	0.2812	0.9596	0.2931	3.4124	1.0421	3.5559	73°40′
16°30′	0.2840	0.9588	0.2962	3.3759	1.0429	3.5209	73°30′
16°40′	0.2868	0.9580	0.2994	3.3402	1.0439	3.4867	73°20′
16°50′	0.2896	0.9572	0.3026	3.3052	1.0448	3.4532	73°10′
17° 0′	0.2924	0.9563	0.3057	3.2709	1.0457	3.4203	73° 0′
17°10′	0.2952	0.9555	0.3089	3.2371	1.0466	3.3881	72°50′
17°20′	0.2979	0.9546	0.3121	3.2041	1.0476	3.3565	72°40′
17°30′	0.3007	0.9537	0.3153	3.1716	1.0485	3.3255	72°30′
17°40′	0.3035	0.9528	0.3185	3.1397	1.0495	3.2951	72°20′
17°50′	0.3062	0.9520	0.3217	3.1084	1.0505	3.2653	72°10′
	cos A	sin A	cot A	tan A	csc A	sec A	A

A	sin A	cos A	tan A	cot A	sec A	csc A	
18° 0′	0.3090	0.9511	0.3249	3.0777	1.0515	3.2361	72° 0′
18°10′	0.3118	0.9502	0.3281	3.0475	1.0525	3.2074	71°50′
18°20′	0.3145	0.9492	0.3314	3.0178	1.0535	3.1792	71°40′
18°30′	0.3173	0.9483	0.3346	2.9887	1.0545	3.1515	71°30′
18°40′	0.3201	0.9474	0.3378	2.9600	1.0555	3.1244	71°20′
18°50′	0.3228	0.9465	0.3411	2.9319	1.0566	3.0977	71°10′
19° 0′	0.3256	0.9455	0.3443	2.9042	1.0576	3.0716	71° 0′
19°10′	0.3283	0.9446	0.3476	2.8770	1.0587	3.0458	70°50′
19°20′	0.3311	0.9436	0.3508	2.8502	1.0598	3.0206	70°40′
19°30′	0.3338	0.9426	0.3541	2.8239	1.0608	2.9957	70°30′
19°40′	0.3365	0.9417	0.3574	2.7980	1.0619	2.9713	70°20′
19°50′	0.3393	0.9407	0.3607	2.7725	1.0631	2.9474	70°10′
20° 0′	0.3420	0.9397	0.3640	2.7475	1.0642	2.9238	70° 0′
20°10′	0.3448	0.9387	0.3673	2.7228	1.0653	2.9006	69°50′
20°20′	0.3475	0.9377	0.3706	2.6985	1.0665	2.8779	69°40′
20°30′	0.3502	0.9367	0.3739	2.6746	1.0676	2.8555	69°30′
20°40′	0.3529	0.9356	0.3772	2.6511	1.0688	2.8334	69°20′
20°50′	0.3557	0.9346	0.3805	2.6279	1.0700	2.8117	69°10′
21° 0′	0.3584	0.9336	0.3839	2.6051	1.0711	2.7904	69° 0′
21°10′	0.3611	0.9325	0.3872	2.5826	1.0723	2.7695	68°50′
21°20′	0.3638	0.9315	0.3906	2.5605	1.0736	2.7488	68°40′
21°30′	0.3665	0.9304	0.3939	2.5386	1.0748	2.7285	68°30′
21°40′	0.3692	0.9293	0.3973	2.5172	1.0760	2.7085	68°20′
21°50′	0.3719	0.9283	0.4006	2.4960	1.0773	2.6888	68°10′
22° 0′	0.3746	0.9272	0.4040	2.4751	1.0785	2.6695	68° 0′
22°10′	0.3773	0.9261	0.4074	2.4545	1.0798	2.6504	67°50′
22°20′	0.3800	0.9250	0.4108	2.4342	1.0811	2.6316	67°40′
22°30′	0.3827	0.9239	0.4142	2.4142	1.0824	2.6131	67°30′
22°40′	0.3854	0.9228	0.4176	2.3945	1.0837	2.5949	67°20′
22°50′	0.3881	0.9216	0.4210	2.3750	1.0850	2.5770	67°10′
23° 0′	0.3907	0.9205	0.4245	2.3559	1.0864	2.5593	67° 0′
23°10′	0.3934	0.9194	0.4279	2.3369	1.0877	2.5419	66°50′
23°20′	0.3961	0.9182	0.4314	2.3183	1.0891	2.5247	66°40′
23°30′	0.3987	0.9171	0.4348	2.2998	1.0904	2.5078	66°30′
23°40′	0.4014	0.9159	0.4383	2.2817	1.0918	2.4912	66°20′
23°50′	0.4041	0.9147	0.4417	2.2637	1.0932	2.4748	66°10′
24° 0′	0.4067	0.9135	0.4452	2.2460	1.0946	2.4586	66° 0′
24°10′	0.4094	0.9124	0.4487	2.2286	1.0961	2.4426	65°50′
24°20′	0.4120	0.9112	0.4522	2.2113	1.0975	2.4269	65°40′
24°30′	0.4147	0.9100	0.4557	2.1943	1.0989	2.4114	65°30′
24°40′	0.4173	0.9088	0.4592	2.1775	1.1004	2.3961	65°20′
24°50′	0.4200	0.9075	0.4628	2.1609	1.1019	2.3811	65°10′
	cos A	sin A	cot A	tan A	csc A	sec A	A

A	sin A	cos A	tan A	cot A	sec A	csc A	
25° 0′	0.4226	0.9063	0.4663	2.1445	1.1034	2.3662	65° 0′
25°10′	0.4253	0.9051	0.4699	2.1283	1.1049	2.3515	64°50′
25°20′	0.4279	0.9038	0.4734	2.1123	1.1064	2.3371	64°40′
25°30′	0.4305	0.9026	0.4770	2.0965	1.1079	2.3228	64°30′
25°40′	0.4331	0.9013	0.4806	2.0809	1.1095	2.3088	64°20′
25°50′	0.4358	0.9001	0.4841	2.0655	1.1110	2.2949	64°10′
26° 0′	0.4384	0.8988	0.4877	2.0503	1.1126	2.2812	64° 0′
26°10′	0.4410	0.8975	0.4913	2.0353	1.1142	2.2677	63°50′
26°20′	0.4436	0.8962	0.4950	2.0204	1.1158	2.2543	63°40′
26°30′	0.4462	0.8949	0.4986	2.0057	1.1174	2.2412	63°30′
26°40′	0.4488	0.8936	0.5022	1.9912	1.1190	2.2282	63°20′
26°50′	0.4514	0.8923	0.5059	1.9768	1.1207	2.2153	63°10′
27°0′	0.4540	0.8910	0.5095	1.9626	1.1223	2.2027	63° 0′
27°10′	0.4566	0.8897	0.5132	1.9486	1.1240	2.1902	62°50′
27°20′	0.4592	0.8884	0.5169	1.9347	1.1257	2.1779	62°40′
27°30′	0.4617	0.8870	0.5206	1.9210	1.1274	2.1657	62°30′
27°40′	0.4643	0.8857	0.5243	1.9074	1.1291	2.1537	62°20′
27°50′	0.4669	0.8843	0.5280	1.8940	1.1308	2.1418	62°10′
28° 0′	0.4695	0.8829	0.5317	1.8807	1.1326	2.1301	62° 0′
28°10′	0.4720	0.8816	0.5354	1.8676	1.1343	2.1185	61°50′
28°20′	0.4746	0.8802	0.5392	1.8546	1.1361	2.1070	61°40′
28°30′	0.4772	0.8788	0.5430	1.8418	1.1379	2.0957	61°30′
28°40′	0.4797	0.8774	0.5467	1.8291	1.1397	2.0846	61°20′
28°50′	0.4823	0.8760	0.5505	1.8165	1.1415	2.0736	61°10′
29° 0′	0.4848	0.8746	0.5543	1.8040	1.1434	2.0627	61° 0′
29°10′	0.4874	0.8732	0.5581	1.7917	1.1452	2.0519	60°50′
29°20′	0.4899	0.8718	0.5619	1.7796	1.1471	2.0413	60°40′
29°30′	0.4924	0.8704	0.5658	1.7675	1.1490	2.0308	60°30′
29°40′	0.4950	0.8689	0.5696	1.7556	1.1509	2.0204	60°20′
29°50′	0.4975	0.8675	0.5735	1.7437	1.1528	2.0101	60°10′
30° 0′	0.5000	0.8660	0.5774	1.7321	1.1547	2.0000	60° 0′
30°10′	0.5025	0.8646	0.5812	1.7205	1.1566	1.9900	59°50′
30°20′	0.5050	0.8631	0.5851	1.7090	1.1586	1.9801	59°40′
30°30′	0.5075	0.8616	0.5890	1.6977	1.1606	1.9703	59°30′
30°40′	0.5100	0.8601	0.5930	1.6864	1.1626	1.9606	59°20′
30°50′	0.5125	0.8587	0.5969	1.6753	1.1646	1.9511	59°10′
31° 0′	0.5150	0.8572	0.6009	1.6643	1.1666	1.9416	59° 0′
31°10′	0.5175	0.8557	0.6048	1.6534	1.1687	1.9323	58°50′
31°20′	0.5200	0.8542	0.6088	1.6426	1.1707	1.9230	58°40′
31°30′	0.5225	0.8526	0.6128	1.6319	1.1728	1.9139	58°30′
31°40′	0.5250	0.8511	0.6168	1.6212	1.1749	1.9048	58°20′
31°50′	0.5275	0.8496	0.6208	1.6107	1.1770	1.8959	58°10′
	cos A	sin A	cot A	tan A	csc A	sec A	A

A	sin A	cos A	tan A	cot A	sec A	csc A	
32° 0′	0.5299	0.8480	0.6249	1.6003	1.1792	1.8871	58° 0′
32°10′	0.5324	0.8465	0.6289	1.5900	1.1813	1.8783	57°50′
32°20′	0.5348	0.8450	0.6330	1.5798	1.1835	1.8697	57°40′
32°30′	0.5373	0.8434	0.6371	1.5697	1.1857	1.8612	57°30′
32°40′	0.5398	0.8418	0.6412	1.5597	1.1879	1.8527	57°20′
32°50′	0.5422	0.8403	0.6453	1.5497	1.1901	1.8443	57°10′
33° 0′	0.5446	0.8387	0.6494	1.5399	1.1924	1.8361	57° 0′
33°10′	0.5471	0.8371	0.6536	1.5301	1.1946	1.8279	56°50′
33°20′	0.5495	0.8355	0.6577	1.5204	1.1969	1.8198	56°40′
33°30′	0.5519	0.8339	0.6619	1.5108	1.1992	1.8118	56°30′
33°40′	0.5544	0.8323	0.6661	1.5013	1.2015	1.8039	56°20′
33°50′	0.5568	0.8307	0.6703	1.4919	1.2039	1.7960	56°10′
34° 0′	0.5592	0.8290	0.6745	1.4826	1.2062	1.7883	56° 0′
34°10′	0.5616	0.8274	0.6787	1.4733	1.2086	1.7806	55°50′
34°20′	0.5640	0.8258	0.6830	1.4641	1.2110	1.7730	55°40′
34°30′	0.5664	0.8241	0.6873	1.4550	1.2134	1.7655	55°30′
34°40′	0.5688	0.8225	0.6916	1.4460	1.2158	1.7581	55°20′
34°50′	0.5712	0.8208	0.6959	1.4370	1.2183	1.7507	55°10′
35° 0′	0.5736	0.8192	0.7002	1.4281	1.2208	1.7434	55° 0′
35°10′	0.5760	0.8175	0.7046	1.4193	1.2233	1.7362	54°50′
35°20′	0.5783	0.8158	0.7089	1.4106	1.2258	1.7291	54°40′
35°30′	0.5807	0.8141	0.7133	1.4019	1.2283	1.7221	54°30′
35°40′	0.5831	0.8124	0.7177	1.3934	1.2309	1.7151	54°20′
35°50′	0.5854	0.8107	0.7221	1.3848	1.2335	1.7081	54°10′
36° 0′	0.5878	0.8090	0.7265	1.3764	1.2361	1.7013	54° 0′
36°10′	0.5901	0.8073	0.7310	1.3680	1.2387	1.6945	53°50′
36°20′	0.5925	0.8056	0.7355	1.3597	1.2413	1.6878	53°40′
36°30′	0.5948	0.8039	0.7400	1.3514	1.2440	1.6812	53°30′
36°40′	0.5972	0.8021	0.7445	1.3432	1.2467	1.6746	53°20′
36°50′	0.5995	0.8004	0.7490	1.3351	1.2494	1.6681	53°10′
37° 0′	0.6018	0.7986	0.7536	1.3270	1.2521	1.6616	53° 0′
37°10′	0.6041	0.7969	0.7581	1.3190	1.2549	1.6553	52°50′
37°20′	0.6065	0.7951	0.7627	1.3111	1.2577	1.6489	52°40′
37°30′	0.6088	0.7934	0.7673	1.3032	1.2605	1.6427	52°30′
37°40′	0.6111	0.7916	0.7720	1.2954	1.2633	1.6365	52°20′
37°50′	0.6134	0.7898	0.7766	1.2876	1.2661	1.6303	52°10′
38° 0′	0.6157	0.7880	0.7813	1.2799	1.2690	1.6243	52° 0′
38°10′	0.6180	0.7862	0.7860	1.2723	1.2719	1.6183	51°50′
38°20′	0.6202	0.7844	0.7907	1.2647	1.2748	1.6123	51°40′
38°30′	0.6225	0.7826	0.7954	1.2572	1.2778	1.6064	51°30′
38°40′	0.6248	0.7808	0.8002	1.2497	1.2807	1.6005	51°20′
38°50′	0.6271	0.7790	0.8050	1.2423	1.2837	1.5948	51°10′
	cos A	sin A	cot A	tan A	csc A	sec A	A

A	sin A	cos A	tan A	cot A	sec A	csc A	
39° 0′	0.6293	0.7771	0.8098	1.2349	1.2868	1.5890	51° 0′
39°10′	0.6316	0.7753	0.8146	1.2276	1.2898	1.5833	50°50′
39°20′	0.6338	0.7735	0.8195	1.2203	1.2929	1.5777	50°40′
39°30′	0.6361	0.7716	0.8243	1.2131	1.2960	1.5721	50°30′
39°40′	0.6383	0.7698	0.8292	1.2059	1.2991	1.5666	50°20′
39°50′	0.6406	0.7679	0.8342	1.1988	1.3022	1.5611	50°10′
40° 0′	0.6428	0.7660	0.8391	1.1918	1.3054	1.5557	50° 0′
40°10′	0.6450	0.7642	0.8441	1.1847	1.3086	1.5504	49°50′
40°20′	0.6472	0.7623	0.8491	1.1778	1.3118	1.5450	49°40′
40°30′	0.6494	0.7604	0.8541	1.1708	1.3151	1.5398	49°30′
40°40′	0.6517	0.7585	0.8591	1.1640	1.3184	1.5345	49°20′
40°50′	0.6539	0.7566	0.8642	1.1571	1.3217	1.5294	49°10′
41° 0′	0.6561	0.7547	0.8693	1.1504	1.3250	1.5243	49° 0′
41°10′	0.6583	0.7528	0.8744	1.1436	1.3284	1.5192	48°50′
41°20′	0.6604	0.7509	0.8796	1.1369	1.3318	1.5141	48°40′
41°30′	0.6626	0.7490	0.8847	1.1303	1.3352	1.5092	48°30′
41°40′	0.6648	0.7470	0.8899	1.1237	1.3386	1.5042	48°20′
41°50′	0.6670	0.7451	0.8952	1.1171	1.3421	1.4993	48°10′
42° 0′	0.6691	0.7431	0.9004	1.1106	1.3456	1.4945	48° 0′
42°10′	0.6713	0.7412	0.9057	1.1041	1.3492	1.4897	47°50′
42°20′	0.6734	0.7392	0.9110	1.0977	1.3527	1.4849	47°40′
42°30′	0.6756	0.7373	0.9163	1.0913	1.3563	1.4802	47°30′
42°40′	0.6777	0.7353	0.9217	1.0850	1.3600	1.4755	47°20′
42°50′	0.6799	0.7333	0.9271	1.0786	1.3636	1.4709	47°10′
43° 0′	0.6820	0.7314	0.9325	1.0724	1.3673	1.4663	47° 0′
43°10′	0.6841	0.7294	0.9380	1.0661	1.3711	1.4617	46°50′
43°20′	0.6862	0.7274	0.9435	1.0599	1.3748	1.4572	46°40′
43°30′	0.6884	0.7254	0.9490	1.0538	1.3786	1.4527	46°30′
43°40′	0.6905	0.7234	0.9545	1.0477	1.3824	1.4483	46°20′
43°50′	0.6926	0.7214	0.9601	1.0416	1.3863	1.4439	46°10′
44° 0′	0.6947	0.7193	0.9657	1.0355	1.3902	1.4396	46° 0′
44°10′	0.6967	0.7173	0.9713	1.0295	1.3941	1.4352	45°50′
44°20′	0.6988	0.7153	0.9770	1.0235	1.3980	1.4310	45°40′
44°30′	0.7009	0.7133	0.9827	1.0176	1.4020	1.4267	45°30′
44°40′	0.7030	0.7112	0.9884	1.0117	1.4061	1.4225	45°20′
44°50′	0.7050	0.7092	0.9942	1.0058	1.4101	1.4183	45°10′
45° 0′	0.7071	0.7071	1.0000	1.0000	1.4142	1.4142	45° 0′
	cos A	sin A	cot A	tan A	csc A	sec A	A

TABLE 2 TRIGONOMETRIC FUNCTIONS—ANGLE IN TENTH OF DEGREE INTERVALS

A	sin A	cos A	tan A	cot A	sec A	csc A	
0.0°	0.0000	1.0000	0.0000	Undefined	1.0000	Undefined	90.0°
0.1°	0.0017	1.0000	0.0017	572.9680	1.0000	572.9590	89.9°
0.2°	0.0035	1.0000	0.0035	286.4750	1.0000	286.4770	89.8°
0.3°	0.0052	1.0000	0.0052	190.9840	1.0000	190.9870	89.7°
0.4°	0.0070	1.0000	0.0070	143.2380	1.0000	143.2410	89.6°
0.5°	0.0087	1.0000	0.0087	114.5880	1.0000	114.5930	89.5°
0.6°	0.0105	0.9999	0.0105	95.4896	1.0001	95.4948	89.4°
0.7°	0.0122	0.9999	0.0122	81.8473	1.0001	81.8534	89.3°
0.8°	0.0140	0.9999	0.0140	71.6150	1.0001	71.6220	89.2°
0.9°	0.0157	0.9999	0.0157	63.6568	1.0001	63.6647	89.1°
1.0°	0.0175	0.9998	0.0175	57.2898	1.0002	57.2986	89.0°
1.1°	0.0192	0.9998	0.0192	52.0806	1.0002	52.0902	88.9°
1.2°	0.0209	0.9998	0.0209	47.7396	1.0002	47.7500	88.8°
1.3°	0.0227	0.9997	0.0227	44.0660	1.0003	44.0774	88.7°
1.4°	0.0244	0.9997	0.0244	40.9174	1.0003	40.9296	88.6°
1.5°	0.0262	0.9997	0.0262	38.1885	1.0003	38.2016	88.5°
1.6°	0.0279	0.9996	0.0279	35.8005	1.0004	35.8145	88.4°
1.7°	0.0297	0.9996	0.0297	33.6935	1.0004	33.7083	88.3°
1.8°	0.0314	0.9995	0.0314	31.8205	1.0005	31.8363	88.2°
1.9°	0.0332	0.9995	0.0332	30.1446	1.0006	30.1612	88.1°
2.0°	0.0349	0.9994	0.0349	28.6363	1.0006	28.6537	88.0°
2.1°	0.0366	0.9993	0.0367	27.2715	1.0007	27.2898	87.9°
2.2°	0.0384	0.9993	0.0384	26.0307	1.0007	26.0499	87.8°
2.3°	0.0401	0.9992	0.0402	24.8978	1.0008	24.9179	87.7°
2.4°	0.0419	0.9991	0.0419	23.8593	1.0009	23.8802	87.6°
2.5°	0.0436	0.9990	0.0437	22.9038	1.0010	22.9256	87.5°
2.6°	0.0454	0.9990	0.0454	22.0217	1.0010	22.0444	87.4°
2.7°	0.0471	0.9989	0.0472	21.2050	1.0011	21.2285	87.3°
2.8°	0.0488	0.9988	0.0489	20.4465	1.0012	20.4709	87.2°
2.9°	0.0506	0.9987	0.0507	19.7403	1.0013	19.7656	87.1°
3.0°	0.0523	0.9986	0.0524	19.0812	1.0014	19.1073	87.0°
3.1°	0.0541	0.9985	0.0542	18.4645	1.0015	18.4915	86.9°
3.2°	0.0558	0.9984	0.0559	17.8863	1.0016	17.9143	86.8°
3.3°	0.0576	0.9983	0.0577	17.3432	1.0017	17.3720	86.7°
3.4°	0.0593	0.9982	0.0594	16.8319	1.0018	16.8616	86.6°
3.5°	0.0610	0.9981	0.0612	16.3499	1.0019	16.3804	86.5°
3.6°	0.0628	0.9980	0.0629	15.8946	1.0020	15.9260	86.4°
3.7°	0.0645	0.9979	0.0647	15.4638	1.0021	15.4961	86.3°
3.8°	0.0663	0.9978	0.0664	15.0557	1.0022	15.0889	86.2°
3.9°	0.0680	0.9977	0.0682	14.6685	1.0023	14.7026	86.1°
4.0°	0.0698	0.9976	0.0699	14.3007	1.0024	14.3356	86.0°
4.1°	0.0715	0.9974	0.0717	13.9507	1.0026	13.9865	85.9°
4.2°	0.0732	0.9973	0.0734	13.6174	1.0027	13.6541	85.8°
	cos A	sin A	cot A	tan A	csc A	sec A	A

A	sin A	cos A	tan A	cot A	sec A	csc A	
4.3°	0.0750	0.9972	0.0752	13.2996	1.0028	13.3371	85.7°
4.4°	0.0767	0.9971	0.0769	12.9962	1.0030	13.0346	85.6°
4.5°	0.0785	0.9969	0.0787	12.7062	1.0031	12.7455	85.5°
4.6°	0.0802	0.9968	0.0805	12.4288	1.0032	12.4690	85.4°
4.7°	0.0819	0.9966	0.0822	12.1632	1.0034	12.2043	85.3°
4.8°	0.0837	0.9965	0.0840	11.9087	1.0035	11.9506	85.2°
4.9°	0.0854	0.9963	0.0857	11.6645	1.0037	11.7073	85.1°
5.0°	0.0872	0.9962	0.0875	11.4301	1.0038	11.4737	85.0°
5.1°	0.0889	0.9960	0.0892	11.2048	1.0040	11.2493	84.9°
5.2°	0.0906	0.9959	0.0910	10.9882	1.0041	11.0336	84.8°
5.3°	0.0924	0.9957	0.0928	10.7797	1.0043	10.8260	84.7°
5.4°	0.0941	0.9956	0.0945	10.5789	1.0045	10.6261	84.6°
5.5°	0.0958	0.9954	0.0963	10.3854	1.0046	10.4334	84.5°
5.6°	0.0976	0.9952	0.0981	10.1988	1.0048	10.2477	84.4°
5.7°	0.0993	0.9951	0.0998	10.0187	1.0050	10.0685	84.3°
5.8°	0.1011	0.9949	0.1016	9.8448	1.0051	9.8955	84.2°
5.9°	0.1028	0.9947	0.1033	9.6768	1.0053	9.7283	84.1°
6.0°	0.1045	0.9945	0.1051	9.5144	1.0055	9.5668	84.0°
6.1°	0.1063	0.9943	0.1069	9.3572	1.0057	9.4105	83.9°
6.2°	0.1080	0.9942	0.1086	9.2052	1.0059	9.2593	83.8°
6.3°	0.1097	0.9940	0.1104	9.0579	1.0061	9.1129	83.7°
6.4°	0.1115	0.9938	0.1122	8.9152	1.0063	8.9711	83.6°
6.5°	0.1132	0.9936	0.1139	8.7769	1.0065	8.8337	83.5°
6.6°	0.1149	0.9934	0.1157	8.6428	1.0067	8.7004	83.4°
6.7°	0.1167	0.9932	0.1175	8.5126	1.0069	8.5711	83.3°
6.8°	0.1184	0.9930	0.1192	8.3863	1.0071	8.4457	83.2°
6.9°	0.1201	0.9928	0.1210	8.2636	1.0073	8.3239	83.1°
7.0°	0.1219	0.9925	0.1228	8.1444	1.0075	8.2055	83.0°
7.1°	0.1236	0.9923	0.1246	8.0285	1.0077	8.0905	82.9°
7.2°	0.1253	0.9921	0.1263	7.9158	1.0079	7.9787	82.8°
7.3°	0.1271	0.9919	0.1281	7.8062	1.0082	7.8700	82.7°
7.4°	0.1288	0.9917	0.1299	7.6996	1.0084	7.7642	82.6°
7.5°	0.1305	0.9914	0.1317	7.5958	1.0086	7.6613	82.5°
7.6°	0.1323	0.9912	0.1334	7.4947	1.0089	7.5611	82.4°
7.7°	0.1340	0.9910	0.1352	7.3962	1.0091	7.4635	82.3°
7.8°	0.1357	0.9907	0.1370	7.3002	1.0093	7.3684	82.2°
7.9°	0.1374	0.9905	0.1388	7.2066	1.0096	7.2757	82.1°
8.0°	0.1392	0.9903	0.1405	7.1154	1.0098	7.1853	82.0°
8.1°	0.1409	0.9900	0.1423	7.0264	1.0101	7.0972	81.9°
8.2°	0.1426	0.9898	0.1441	6.9395	1.0103	7.0112	81.8°
8.3°	0.1444	0.9895	0.1459	6.8548	1.0106	6.9273	81.7°
8.4°	0.1461	0.9893	0.1477	6.7720	1.0108	6.8454	81.6°
8.5°	0.1478	0.9890	0.1495	6.6912	1.0111	6.7655	81.5°
8.6°	0.1495	0.9888	0.1512	6.6122	1.0114	6.6874	81.4°
8.7°	0.1513	0.9885	0.1530	6.5350	1.0116	6.6111	81.3°
	cos A	sin A	cot A	tan A	csc A	sec A	A

A	sin A	cos A	tan A	cot A	sec A	csc A	
8.8°	0.1530	0.9882	0.1548	6.4596	1.0119	6.5366	81.2°
8.9°	0.1547	0.9880	0.1566	6.3859	1.0122	6.4637	81.1°
9.0°	0.1564	0.9877	0.1584	6.3138	1.0125	6.3925	81.0°
9.1°	0.1582	0.9874	0.1602	6.2432	1.0127	6.3228	80.9°
9.2°	0.1599	0.9871	0.1620	6.1742	1.0130	6.2546	80.8°
9.3°	0.1616	0.9869	0.1638	6.1066	1.0133	6.1880	80.7°
9.4°	0.1633	0.9866	0.1655	6.0405	1.0136	6.1227	80.6°
9.5°	0.1650	0.9863	0.1673	5.9758	1.0139	6.0589	80.5°
9.6°	0.1668	0.9860	0.1691	5.9124	1.0142	5.9963	80.4°
9.7°	0.1685	0.9857	0.1709	5.8502	1.0145	5.9351	80.3°
9.8°	0.1702	0.9854	0.1727	5.7894	1.0148	5.8751	80.2°
9.9°	0.1719	0.9851	0.1745	5.7297	1.0151	5.8164	80.1°
10.0°	0.1736	0.9848	0.1763	5.6713	1.0154	5.7588	80.0°
10.1°	0.1754	0.9845	0.1781	5.6140	1.0157	5.7023	79.9°
10.2°	0.1771	0.9842	0.1799	5.5578	1.0161	5.6470	79.8°
10.3°	0.1788	0.9839	0.1817	5.5026	1.0164	5.5928	79.7°
10.4°	0.1805	0.9836	0.1835	5.4486	1.0167	5.5396	79.6°
10.5°	0.1822	0.9833	0.1853	5.3955	1.0170	5.4874	79.5°
10.6°	0.1840	0.9829	0.1871	5.3435	1.0174	5.4362	79.4°
10.7°	0.1857	0.9826	0.1890	5.2923	1.0177	5.3860	79.3°
10.8°	0.1874	0.9823	0.1908	5.2422	1.0180	5.3367	79.2°
10.9°	0.1891	0.9820	0.1926	5.1929	1.0184	5.2883	79.1°
11.0°	0.1908	0.9816	0.1944	5.1446	1.0187	5.2408	79.0°
11.1°	0.1925	0.9813	0.1962	5.0970	1.0191	5.1942	78.9°
11.2°	0.1942	0.9810	0.1980	5.0504	1.0194	5.1484	78.8°
11.3°	0.1959	0.9806	0.1998	5.0045	1.0198	5.1034	78.7°
11.4°	0.1977	0.9803	0.2016	4.9594	1.0201	5.0593	78.6°
11.5°	0.1994	0.9799	0.2035	4.9152	1.0205	5.0158	78.5°
11.6°	0.2011	0.9796	0.2053	4.8716	1.0209	4.9732	78.4°
11.7°	0.2028	0.9792	0.2071	4.8288	1.0212	4.9313	78.3°
11.8°	0.2045	0.9789	0.2089	4.7867	1.0216	4.8901	78.2°
11.9°	0.2062	0.9785	0.2107	4.7453	1.0220	4.8496	78.1°
12.0°	0.2079	0.9781	0.2126	4.7046	1.0223	4.8097	78.0°
12.1°	0.2096	0.9778	0.2144	4.6646	1.0227	4.7706	77.9°
12.2°	0.2113	0.9774	0.2162	4.6252	1.0231	4.7320	77.8°
12.3°	0.2130	0.9770	0.2180	4.5864	1.0235	4.6942	77.7°
12.4°	0.2147	0.9767	0.2199	4.5483	1.0239	4.6569	77.6°
12.5°	0.2164	0.9763	0.2217	4.5107	1.0243	4.6202	77.5°
12.6°	0.2181	0.9759	0.2235	4.4737	1.0247	4.5841	77.4°
12.7°	0.2198	0.9755	0.2254	4.4373	1.0251	4.5486	77.3°
12.8°	0.2215	0.9751	0.2272	4.4015	1.0255	4.5137	77.2°
12.9°	0.2233	0.9748	0.2290	4.3662	1.0259	4.4793	77.1°
13.0°	0.2250	0.9744	0.2309	4.3315	1.0263	4.4454	77.0°
13.1°	0.2267	0.9740	0.2327	4.2972	1.0267	4.4121	76.9°
	cos A	sin A	cot A	tan A	csc A	sec A	A

A	sin A	cos A	tan A	cot A	sec A	csc A	
13.2°	0.2284	0.9736	0.2345	4.2635	1.0271	4.3792	76.8°
13.3°	0.2300	0.9732	0.2364	4.2303	1.0276	4.3469	76.7°
13.4°	0.2317	0.9728	0.2382	4.1976	1.0280	4.3150	76.6°
13.5°	0.2334	0.9724	0.2401	4.1653	1.0284	4.2837	76.5°
13.6°	0.2351	0.9720	0.2419	4.1335	1.0288	4.2527	76.4°
13.7°	0.2368	0.9715	0.2438	4.1022	1.0293	4.2223	76.3°
13.8°	0.2385	0.9711	0.2456	4.0713	1.0297	4.1923	76.2°
13.9°	0.2402	0.9707	0.2475	4.0408	1.0302	4.1627	76.1°
14.0°	0.2419	0.9703	0.2493	4.0108	1.0306	4.1336	76.0°
14.1°	0.2436	0.9699	0.2512	3.9812	1.0311	4.1048	75.9°
14.2°	0.2453	0.9694	0.2530	3.9520	1.0315	4.0765	75.8°
14.3°	0.2470	0.9690	0.2549	3.9232	1.0320	4.0486	75.7°
14.4°	0.2487	0.9686	0.2568	3.8947	1.0324	4.0211	75.6°
14.5°	0.2504	0.9681	0.2586	3.8667	1.0329	3.9939	75.5°
14.6°	0.2521	0.9677	0.2605	3.8391	1.0334	3.9672	75.4°
14.7°	0.2538	0.9673	0.2623	3.8118	1.0338	3.9408	75.3°
14.8°	0.2554	0.9668	0.2642	3.7848	1.0343	3.9147	75.2°
14.9°	0.2571	0.9664	0.2661	3.7583	1.0348	3.8890	75.1°
15.0°	0.2588	0.9659	0.2679	3.7320	1.0353	3.8637	75.0°
15.1°	0.2605	0.9655	0.2698	3.7062	1.0358	3.8387	74.9°
15.2°	0.2622	0.9650	0.2717	3.6806	1.0363	3.8140	74.8°
15.3°	0.2639	0.9646	0.2736	3.6554	1.0367	3.7897	74.7°
15.4°	0.2656	0.9641	0.2754	3.6305	1.0372	3.7657	74.6°
15.5°	0.2672	0.9636	0.2773	3.6059	1.0377	3.7420	74.5°
15.6°	0.2689	0.9632	0.2792	3.5816	1.0382	3.7186	74.4°
15.7°	0.2706	0.9627	0.2811	3.5576	1.0388	3.6955	74.3°
15.8°	0.2723	0.9622	0.2830	3.5339	1.0393	3.6727	74.2°
15.9°	0.2740	0.9617	0.2849	3.5105	1.0398	3.6502	74.1°
16.0°	0.2756	0.9613	0.2867	3.4874	1.0403	3.6280	74.0°
16.1°	0.2773	0.9608	0.2886	3.4646	1.0408	3.6060	73.9°
16.2°	0.2790	0.9603	0.2905	3.4420	1.0413	3.5843	73.8°
16.3°	0.2807	0.9598	0.2924	3.4197	1.0419	3.5629	73.7°
16.4°	0.2823	0.9593	0.2943	3.3977	1.0424	3.5418	73.6°
16.5°	0.2840	0.9588	0.2962	3.3759	1.0429	3.5209	73.5°
16.6°	0.2857	0.9583	0.2981	3.3544	1.0435	3.5003	73.4°
16.7°	0.2874	0.9578	0.3000	3.3332	1.0440	3.4799	73.3°
16.8°	0.2890	0.9573	0.3019	3.3122	1.0446	3.4598	73.2°
16.9°	0.2907	0.9568	0.3038	3.2914	1.0451	3.4399	73.1°
17.0°	0.2924	0.9563	0.3057	3.2708	1.0457	3.4203	73.0°
17.1°	0.2940	0.9558	0.3076	3.2505	1.0463	3.4009	72.9°
17.2°	0.2957	0.9553	0.3096	3.2305	1.0468	3.3817	72.8°
17.3°	0.2974	0.9548	0.3115	3.2106	1.0474	3.3628	72.7°
17.4°	0.2990	0.9542	0.3134	3.1910	1.0480	3.3440	72.6°
17.5°	0.3007	0.9537	0.3153	3.1716	1.0485	3.3255	72.5°
17.6°	0.3024	0.9532	0.3172	3.1524	1.0491	3.3072	72.4°
	cos A	sin A	cot A	tan A	csc A	sec A	A

A	sin A	cos A	tan A	cot A	sec A	csc A	
17.7°	0.3040	0.9527	0.3191	3.1334	1.0497	3.2891	72.3°
17.8°	0.3057	0.9521	0.3211	3.1146	1.0503	3.2712	72.2°
17.9°	0.3074	0.9516	0.3230	3.0961	1.0509	3.2535	72.1°
18.0°	0.3090	0.9511	0.3249	3.0777	1.0515	3.2361	72.0°
18.1°	0.3107	0.9505	0.3269	3.0595	1.0521	3.2188	71.9°
18.2°	0.3123	0.9500	0.3288	3.0415	1.0527	3.2017	71.8°
18.3°	0.3140	0.9494	0.3307	3.0237	1.0533	3.1848	71.7°
18.4°	0.3156	0.9489	0.3327	3.0061	1.0539	3.1681	71.6°
18.5°	0.3173	0.9483	0.3346	2.9887	1.0545	3.1515	71.5°
18.6°	0.3190	0.9478	0.3365	2.9714	1.0551	3.1352	71.4°
18.7°	0.3206	0.9472	0.3385	2.9544	1.0557	3.1190	71.3°
18.8°	0.3223	0.9466	0.3404	2.9375	1.0564	3.1030	71.2°
18.9°	0.3239	0.9461	0.3424	2.9208	1.0570	3.0872	71.1°
19.0°	0.3256	0.9455	0.3443	2.9042	1.0576	3.0715	71.0°
19.1°	0.3272	0.9449	0.3463	2.8878	1.0583	3.0561	70.9°
19.2°	0.3289	0.9444	0.3482	2.8716	1.0589	3.0407	70.8°
19.3°	0.3305	0.9438	0.3502	2.8555	1.0595	3.0256	70.7°
19.4°	0.3322	0.9432	0.3522	2.8396	1.0602	3.0106	70.6°
19.5°	0.3338	0.9426	0.3541	2.8239	1.0608	2.9957	70.5°
19.6°	0.3355	0.9421	0.3561	2.8083	1.0615	2.9811	70.4°
19.7°	0.3371	0.9415	0.3581	2.7929	1.0622	2.9665	70.3°
19.8°	0.3387	0.9409	0.3600	2.7776	1.0628	2.9521	70.2°
19.9°	0.3404	0.9403	0.3620	2.7625	1.0635	2.9379	70.1°
20.0°	0.3420	0.9397	0.3640	2.7475	1.0642	2.9238	70.0°
20.1°	0.3437	0.9391	0.3659	2.7326	1.0649	2.9098	69.9°
20.2°	0.3453	0.9385	0.3679	2.7179	1.0655	2.8960	69.8°
20.3°	0.3469	0.9379	0.3699	2.7033	1.0662	2.8824	69.7°
20.4°	0.3486	0.9373	0.3719	2.6889	1.0669	2.8688	69.6°
20.5°	0.3502	0.9367	0.3739	2.6746	1.0676	2.8554	69.5°
20.6°	0.3518	0.9361	0.3759	2.6605	1.0683	2.8422	69.4°
20.7°	0.3535	0.9354	0.3779	2.6464	1.0690	2.8291	69.3°
20.8°	0.3551	0.9348	0.3799	2.6325	1.0697	2.8160	69.2°
20.9°	0.3567	0.9342	0.3819	2.6187	1.0704	2.8032	69.1°
21.0°	0.3584	0.9336	0.3839	2.6051	1.0711	2.7904	69.0°
21.1°	0.3600	0.9330	0.3859	2.5916	1.0719	2.7778	68.9°
21.2°	0.3616	0.9323	0.3879	2.5781	1.0726	2.7653	68.8°
21.3°	0.3633	0.9317	0.3899	2.5649	1.0733	2.7529	68.7°
21.4°	0.3649	0.9311	0.3919	2.5517	1.0740	2.7406	68.6°
21.5°	0.3665	0.9304	0.3939	2.5386	1.0748	2.7285	68.5°
21.6°	0.3681	0.9298	0.3959	2.5257	1.0755	2.7165	68.4°
21.7°	0.3697	0.9291	0.3979	2.5129	1.0763	2.7045	68.3°
21.8°	0.3714	0.9285	0.4000	2.5002	1.0770	2.6927	68.2°
21.9°	0.3730	0.9278	0.4020	2.4876	1.0778	2.6810	68.1°
22.0°	0.3746	0.9272	0.4040	2.4751	1.0785	2.6695	68.0°
	cos A	sin A	cot A	tan A	csc A	sec A	A

A	sin A	cos A	tan A	cot A	sec A	csc A	
22.1°	0.3762	0.9265	0.4061	2.4627	1.0793	2.6580	67.9°
22.2°	0.3778	0.9259	0.4081	2.4504	1.0801	2.6466	67.8°
22.3°	0.3795	0.9252	0.4101	2.4382	1.0808	2.6353	67.7°
22.4°	0.3811	0.9245	0.4122	2.4262	1.0816	2.6242	67.6°
22.5°	0.3827	0.9239	0.4142	2.4142	1.0824	2.6131	67.5°
22.6°	0.3843	0.9232	0.4163	2.4023	1.0832	2.6022	67.4°
22.7°	0.3859	0.9225	0.4183	2.3906	1.0840	2.5913	67.3°
22.8°	0.3875	0.9219	0.4204	2.3789	1.0848	2.5805	67.2°
22.9°	0.3891	0.9212	0.4224	2.3673	1.0856	2.5699	67.1°
23.0°	0.3907	0.9205	0.4245	2.3558	1.0864	2.5593	67.0°
23.1°	0.3923	0.9198	0.4265	2.3445	1.0872	2.5488	66.9°
23.2°	0.3939	0.9191	0.4286	2.3332	1.0880	2.5384	66.8°
23.3°	0.3955	0.9184	0.4307	2.3220	1.0888	2.5281	66.7°
23.4°	0.3971	0.9178	0.4327	2.3109	1.0896	2.5179	66.6°
23.5°	0.3987	0.9171	0.4348	2.2998	1.0904	2.5078	66.5°
23.6°	0.4003	0.9164	0.4369	2.2889	1.0913	2.4978	66.4°
23.7°	0.4019	0.9157	0.4390	2.2781	1.0921	2.4879	66.3°
23.8°	0.4035	0.9150	0.4411	2.2673	1.0929	2.4780	66.2°
23.9°	0.4051	0.9143	0.4431	2.2566	1.0938	2.4683	66.1°
24.0°	0.4067	0.9135	0.4452	2.2460	1.0946	2.4586	66.0°
24.1°	0.4083	0.9128	0.4473	2.2355	1.0955	2.4490	65.9°
24.2°	0.4099	0.9121	0.4494	2.2251	1.0963	2.4395	65.8°
24.3°	0.4115	0.9114	0.4515	2.2147	1.0972	2.4300	65.7°
24.4°	0.4131	0.9107	0.4536	2.2045	1.0981	2.4207	65.6°
24.5°	0.4147	0.9100	0.4557	2.1943	1.0989	2.4114	65.5°
24.6°	0.4163	0.9092	0.4578	2.1842	1.0998	2.4022	65.4°
24.7°	0.4179	0.9085	0.4599	2.1742	1.1007	2.3931	65.3°
24.8°	0.4195	0.9078	0.4621	2.1642	1.1016	2.3841	65.2°
24.9°	0.4210	0.9070	0.4642	2.1543	1.1025	2.3751	65.1°
25.0°	0.4226	0.9063	0.4663	2.1445	1.1034	2.3662	65.0°
25.1°	0.4242	0.9056	0.4684	2.1348	1.1043	2.3574	64.9°
25.2°	0.4258	0.9048	0.4706	2.1251	1.1052	2.3486	64.8°
25.3°	0.4274	0.9041	0.4727	2.1155	1.1061	2.3400	64.7°
25.4°	0.4289	0.9033	0.4748	2.1060	1.1070	2.3313	64.6°
25.5°	0.4305	0.9026	0.4770	2.0965	1.1079	2.3228	64.5°
25.6°	0.4321	0.9018	0.4791	2.0872	1.1089	2.3144	64.4°
25.7°	0.4337	0.9011	0.4813	2.0778	1.1098	2.3060	64.3°
25.8°	0.4352	0.9003	0.4834	2.0686	1.1107	2.2976	64.2°
25.9°	0.4368	0.8996	0.4856	2.0594	1.1117	2.2894	64.1°
26.0°	0.4384	0.8988	0.4877	2.0503	1.1126	2.2812	64.0°
26.1°	0.4399	0.8980	0.4899	2.0412	1.1136	2.2730	63.9°
26.2°	0.4415	0.8973	0.4921	2.0323	1.1145	2.2650	63.8°
26.3°	0.4431	0.8965	0.4942	2.0233	1.1155	2.2570	63.7°
26.4°	0.4446	0.8957	0.4964	2.0145	1.1164	2.2490	63.6°
26.5°	0.4462	0.8949	0.4986	2.0057	1.1174	2.2412	63.5°
	cos A	sin A	cot A	tan A	csc A	sec A	A

A	sin A	cos A	tan A	cot A	sec A	csc A	
26.6°	0.4478	0.8942	0.5008	1.9969	1.1184	2.2333	63.4°
26.7°	0.4493	0.8934	0.5029	1.9883	1.1194	2.2256	63.3°
26.8°	0.4509	0.8926	0.5051	1.9797	1.1203	2.2179	63.2°
26.9°	0.4524	0.8918	0.5073	1.9711	1.1213	2.2103	63.1°
27.0°	0.4540	0.8910	0.5095	1.9626	1.1223	2.2027	63.0°
27.1°	0.4555	0.8902	0.5117	1.9542	1.1233	2.1952	62.9°
27.2°	0.4571	0.8894	0.5139	1.9458	1.1243	2.1877	62.8°
27.3°	0.4587	0.8886	0.5161	1.9375	1.1253	2.1803	62.7°
27.4°	0.4602	0.8878	0.5184	1.9292	1.1264	2.1730	62.6°
27.5°	0.4617	0.8870	0.5206	1.9210	1.1274	2.1657	62.5°
27.6°	0.4633	0.8862	0.5228	1.9128	1.1284	2.1584	62.4°
27.7°	0.4648	0.8854	0.5250	1.9047	1.1294	2.1513	62.3°
27.8°	0.4664	0.8846	0.5272	1.8967	1.1305	2.1441	62.2°
27.9°	0.4679	0.8838	0.5295	1.8887	1.1315	2.1371	62.1°
28.0°	0.4695	0.8829	0.5347	1.8807	1.1326	2.1300	62.0°
28.1°	0.4710	0.8821	0.5340	1.8728	1.1336	2.1231	61.9°
28.2°	0.4726	0.8813	0.5362	1.8650	1.1347	2.1162	61.8°
28.3°	0.4741	0.8805	0.5384	1.8572	1.1357	2.1093	61.7°
28.4°	0.4756	0.8796	0.5407	1.8495	1.1368	2.1025	61.6°
28.5°	0.4772	0.8788	0.5430	1.8418	1.1379	2.0957	61.5°
28.6°	0.4787	0.8780	0.5452	1.8341	1.1390	2.0890	61.4°
28.7°	0.4802	0.8771	0.5475	1.8265	1.1401	2.0824	61.3°
28.8°	0.4818	0.8763	0.5498	1.8190	1.1412	2.0757	61.2°
28.9°	0.4833	0.8755	0.5520	1.8115	1.1423	2.0692	61.1°
29.0°	0.4848	0.8746	0.5543	1.8040	1.1434	2.0627	61.0°
29.1°	0.4863	0.8738	0.5566	1.7966	1.1445	2.0562	60.9°
29.2°	0.4879	0.8729	0.5589	1.7893	1.1456	2.0498	60.8°
29.3°	0.4894	0.8721	0.5612	1.7820	1.1467	2.0434	60.7°
29.4°	0.4909	0.8712	0.5635	1.7747	1.1478	2.0371	60.6°
29.5°	0.4924	0.8704	0.5658	1.7675	1.1490	2.0308	60.5°
29.6°	0.4939	0.8695	0.5681	1.7603	1.1501	2.0245	60.4°
29.7°	0.4955	0.8686	0.5704	1.7532	1.1512	2.0183	60.3°
29.8°	0.4970	0.8678	0.5727	1.7461	1.1524	2.0122	60.2°
29.9°	0.4985	0.8669	0.5750	1.7390	1.1535	2.0061	60.1°
30.0°	0.5000	0.8660	0.5774	1.7320	1.1547	2.0000	60.0°
30.1°	0.5015	0.8652	0.5797	1.7251	1.1559	1.9940	59.9°
30.2°	0.5030	0.8643	0.5820	1.7182	1.1570	1.9880	59.8°
30.3°	0.5045	0.8634	0.5844	1.7113	1.1582	1.9820	59.7°
30.4°	0.5060	0.8625	0.5867	1.7045	1.1594	1.9761	59.6°
30.5°	0.5075	0.8616	0.5890	1.6977	1.1606	1.9703	59.5°
30.6°	0.5090	0.8607	0.5914	1.6909	1.1618	1.9645	59.4°
30.7°	0.5105	0.8599	0.5938	1.6842	1.1630	1.9587	59.3°
30.8°	0.5120	0.8590	0.5961	1.6775	1.1642	1.9530	59.2°
30.9°	0.5135	0.8581	0.5985	1.6709	1.1654	1.9473	59.1°
	cos A	sin A	cot A	tan A	csc A	sec A	A

A	sin A	cos A	tan A	cot A	sec A	csc A	
31.0°	0.5150	0.8572	0.6009	1.6643	1.1666	1.9416	59.0°
31.1°	0.5165	0.8563	0.6032	1.6577	1.1679	1.9360	58.9°
31.2°	0.5180	0.8554	0.6056	1.6512	1.1691	1.9304	58.8°
31.3°	0.5195	0.8545	0.6080	1.6447	1.1703	1.9249	58.7°
31.4°	0.5210	0.8536	0.6104	1.6383	1.1716	1.9193	58.6°
31.5°	0.5225	0.8526	0.6128	1.6318	1.1728	1.9139	58.5°
31.6°	0.5240	0.8517	0.6152	1.6255	1.1741	1.9084	58.4°
31.7°	0.5255	0.8508	0.6176	1.6191	1.1754	1.9030	58.3°
31.8°	0.5270	0.8499	0.6200	1.6128	1.1766	1.8977	58.2°
31.9°	0.5284	0.8490	0.6224	1.6066	1.1779	1.8924	58.1°
32.0°	0.5299	0.8480	0.6249	1.6003	1.1792	1.8871	58.0°
32.1°	0.5314	0.8471	0.6273	1.5941	1.1805	1.8818	57.9°
32.2°	0.5329	0.8462	0.6297	1.5880	1.1818	1.8766	57.8°
32.3°	0.5344	0.8453	0.6322	1.5818	1.1831	1.8714	57.7°
32.4°	0.5358	0.8443	0.6346	1.5757	1.1844	1.8663	57.6°
32.5°	0.5373	0.8434	0.6371	1.5697	1.1857	1.8612	57.5°
32.6°	0.5388	0.8425	0.6395	1.5637	1.1870	1.8561	57.4°
32.7°	0.5402	0.8415	0.6420	1.5577	1.1883	1.8510	57.3°
32.8°	0.5417	0.8406	0.6445	1.5517	1.1897	1.8460	57.2°
32.9°	0.5432	0.8396	0.6469	1.5458	1.1910	1.8410	57.1°
33.0°	0.5446	0.8387	0.6494	1.5399	1.1924	1.8361	57.0°
33.1°	0.5461	0.8377	0.6519	1.5340	1.1937	1.8312	56.9°
33.2°	0.5476	0.8368	0.6544	1.5282	1.1951	1.8263	56.8°
33.3°	0.5490	0.8358	0.6569	1.5224	1.1964	1.8214	56.7°
33.4°	0.5505	0.8348	0.6594	1.5166	1.1978	1.8166	56.6°
33.5°	0.5519	0.8339	0.6619	1.5108	1.1992	1.8118	56.5°
33.6°	0.5534	0.8329	0.6644	1.5051	1.2006	1.8070	56.4°
33.7°	0.5548	0.8320	0.6669	1.4994	1.2020	1.8023	56.3°
33.8°	0.5563	0.8310	0.6694	1.4938	1.2034	1.7976	56.2°
33.9°	0.5577	0.8300	0.6720	1.4882	1.2048	1.7929	56.1°
34.0°	0.5592	0.8290	0.6745	1.4826	1.2062	1.7883	56.0°
34.1°	0.5606	0.8281	0.6771	1.4770	1.2076	1.7837	55.9°
34.2°	0.5621	0.8271	0.6796	1.4715	1.2091	1.7791	58.8°
34.3°	0.5635	0.8261	0.6822	1.4659	1.2105	1.7745	55.7°
34.4°	0.5650	0.8251	0.6847	1.4605	1.2120	1.7700	55.6°
34.5°	0.5664	0.8241	0.6873	1.4550	1.2134	1.7655	55.5°
34.6°	0.5678	0.8231	0.6899	1.4496	1.2149	1.7610	55.4°
34.7°	0.5693	0.8221	0.6924	1.4442	1.2163	1.7566	55.3°
34.8°	0.5707	0.8211	0.6950	1.4388	1.2178	1.7522	55.2°
34.9°	0.5721	0.8202	0.6976	1.4335	1.2193	1.7478	55.1°
35.0°	0.5736	0.8192	0.7002	1.4281	1.2208	1.7434	55.0°
35.1°	0.5750	0.8181	0.7028	1.4229	1.2223	1.7391	54.9°
35.2°	0.5764	0.8171	0.7054	1.4176	1.2238	1.7348	54.8°
35.3°	0.5779	0.8161	0.7080	1.4123	1.2253	1.7305	54.7°
35.4°	0.5793	0.8151	0.7107	1.4071	1.2268	1.7263	54.6°
	cos A	sin A	cot A	tan A	csc A	sec A	A

A	sin A	cos A	tan A	cot A	sec A	csc A	
35.5°	0.5807	0.8141	0.7133	1.4019	1.2283	1.7220	54.5°
35.6°	0.5821	0.8131	0.7159	1.3968	1.2299	1.7178	54.4°
35.7°	0.5835	0.8121	0.7186	1.3916	1.2314	1.7137	54.3°
35.8°	0.5850	0.8111	0.7212	1.3865	1.2329	1.7095	54.2°
35.9°	0.5864	0.8100	0.7239	1.3814	1.2345	1.7054	54.1°
36.0°	0.5878	0.8090	0.7265	1.3764	1.2361	1.7013	54.0°
36.1°	0.5892	0.8080	0.7292	1.3713	1.2376	1.6972	53.9°
36.2°	0.5906	0.8070	0.7319	1.3663	1.2392	1.6932	53.8°
36.3°	0.5920	0.8059	0.7346	1.3613	1.2408	1.6892	53.7°
36.4°	0.5934	0.8049	0.7373	1.3564	1.2424	1.6851	53.6°
36.5°	0.5948	0.8039	0.7400	1.3514	1.2440	1.6812	53.5°
36.6°	0.5962	0.8028	0.7427	1.3465	1.2456	1.6772	53.4°
36.7°	0.5976	0.8018	0.7454	1.3416	1.2472	1.6733	53.3°
36.8°	0.5990	0.8007	0.7481	1.3367	1.2489	1.6694	53.2°
36.9°	0.6004	0.7997	0.7508	1.3319	1.2505	1.6655	53.1°
37.0°	0.6018	0.7986	0.7536	1.3270	1.2521	1.6616	53.0°
37.1°	0.6032	0.7976	0.7563	1.3222	1.2538	1.6578	52.9°
37.2°	0.6046	0.7965	0.7590	1.3175	1.2554	1.6540	52.8°
37.3°	0.6060	0.7955	0.7618	1.3127	1.2571	1.6502	52.7°
37.4°	0.6074	0.7944	0.7646	1.3079	1.2588	1.6464	52.6°
37.5°	0.6088	0.7934	0.7673	1.3032	1.2605	1.6427	52.5°
37.6°	0.6101	0.7923	0.7701	1.2985	1.2622	1.6390	52.4°
37.7°	0.6115	0.7912	0.7729	1.2938	1.2639	1.6353	52.3°
37.8°	0.6129	0.7902	0.7757	1.2892	1.2656	1.6316	52.2°
37.9°	0.6143	0.7891	0.7785	1.2846	1.2673	1.6279	52.1°
38.0°	0.6157	0.7880	0.7813	1.2799	1.2690	1.6243	52.0°
38.1°	0.6170	0.7869	0.7841	1.2753	1.2708	1.6207	51.9°
38.2°	0.6184	0.7859	0.7869	1.2708	1.2725	1.6171	51.8°
38.3°	0.6198	0.7848	0.7898	1.2662	1.2742	1.6135	51.7°
38.4°	0.6211	0.7837	0.7926	1.2617	1.2760	1.6099	51.6°
38.5°	0.6225	0.7826	0.7954	1.2572	1.2778	1.6064	51.5°
38.6°	0.6239	0.7815	0.7983	1.2527	1.2796	1.6029	51.4°
38.7°	0.6252	0.7804	0.8012	1.2482	1.2813	1.5994	51.3°
38.8°	0.6266	0.7793	0.8040	1.2438	1.2831	1.5959	51.2°
38.9°	0.6280	0.7782	0.8069	1.2393	1.2849	1.5925	51.1°
39.0°	0.6293	0.7771	0.8098	1.2349	1.2868	1.5890	51.0°
39.1°	0.6307	0.7760	0.8127	1.2305	1.2886	1.5856	50.9°
39.2°	0.6320	0.7749	0.8156	1.2261	1.2904	1.5822	50.8°
39.3°	0.6334	0.7738	0.8185	1.2218	1.2923	1.5788	50.7°
39.4°	0.6347	0.7727	0.8214	1.2174	1.2941	1.5755	50.6°
39.5°	0.6361	0.7716	0.8243	1.2131	1.2960	1.5721	50.5°
39.6°	0.6374	0.7705	0.8273	1.2088	1.2978	1.5688	50.4°
39.7°	0.6388	0.7694	0.8302	1.2045	1.2997	1.5655	50.3°
39.8°	0.6401	0.7683	0.8332	1.2002	1.3016	1.5622	50.2°
39.9°	0.6414	0.7672	0.8361	1.1960	1.3035	1.5590	50.1°
	cos A	sin A	cot A	tan A	csc A	sec A	A

A	sin A	cos A	tan A	cot A	sec A	csc A	
40.0°	0.6428	0.7660	0.8391	1.1918	1.3054	1.5557	50.0°
40.1°	0.6441	0.7649	0.8421	1.1875	1.3073	1.5525	49.9°
40.2°	0.6455	0.7638	0.8451	1.1833	1.3092	1.5493	49.8°
40.3°	0.6468	0.7627	0.8481	1.1792	1.3112	1.5461	49.7°
40.4°	0.6481	0.7615	0.8511	1.1750	1.3131	1.5429	49.6°
40.5°	0.6494	0.7604	0.8541	1.1709	1.3151	1.5398	49.5°
40.6°	0.6508	0.7593	0.8571	1.1667	1.3171	1.5366	49.4°
40.7°	0.6521	0.7581	0.8601	1.1626	1.3190	1.5335	49.3°
40.8°	0.6534	0.7570	0.8632	1.1585	1.3210	1.5304	49.2°
40.9°	0.6547	0.7559	0.8662	1.1544	1.3230	1.5273	49.1°
41.0°	0.6561	0.7547	0.8693	1.1504	1.3250	1.5243	49.0°
41.1°	0.6574	0.7536	0.8724	1.1463	1.3270	1.5212	48.9°
41.2°	0.6587	0.7524	0.8754	1.1423	1.3291	1.5182	48.8°
41.3°	0.6600	0.7513	0.8785	1.1383	1.3311	1.5151	48.7°
41.4°	0.6613	0.7501	0.8816	1.1343	1.3331	1.5121	48.6°
41.5°	0.6626	0.7490	0.8847	1.1303	1.3352	1.5092	48.5°
41.6°	0.6639	0.7478	0.8878	1.1263	1.3373	1.5062	48.4°
41.7°	0.6652	0.7466	0.8910	1.1224	1.3393	1.5032	48.3°
41.8°	0.6665	0.7455	0.8941	1.1184	1.3414	1.5003	48.2°
41.9°	0.6678	0.7443	0.8972	1.1145	1.3435	1.4974	48.1°
42.0°	0.6691	0.7431	0.9004	1.1106	1.3456	1.4945	48.0°
42.1°	0.6704	0.7420	0.9036	1.1067	1.3478	1.4916	47.9°
42.2°	0.6717	0.7408	0.9067	1.1028	1.3499	1.4887	47.8°
42.3°	0.6730	0.7396	0.9099	1.0990	1.3520	1.4859	47.7°
42.4°	0.6743	0.7385	0.9131	1.0951	1.3542	1.4830	47.6°
42.5°	0.6756	0.7373	0.9163	1.0913	1.3563	1.4802	47.5°
42.6°	0.6769	0.7361	0.9195	1.0875	1.3585	1.4774	47.4°
42.7°	0.6782	0.7349	0.9228	1.0837	1.3607	1.4746	47.3°
42.8°	0.6794	0.7337	0.9260	1.0799	1.3629	1.4718	47.2°
42.9°	0.6807	0.7325	0.9293	1.0761	1.3651	1.4690	47.1°
43.0°	0.6820	0.7314	0.9325	1.0724	1.3673	1.4663	47.0°
43.1°	0.6833	0.7302	0.9358	1.0686	1.3696	1.4635	46.9°
43.2°	0.6845	0.7290	0.9391	1.0649	1.3718	1.4608	46.8°
43.3°	0.6858	0.7278	0.9423	1.0612	1.3741	1.4581	46.7°
43.4°	0.6871	0.7266	0.9457	1.0575	1.3763	1.4554	46.6°
43.5°	0.6884	0.7254	0.9490	1.0538	1.3786	1.4527	46.5°
43.6°	0.6896	0.7242	0.9523	1.0501	1.3809	1.4501	46.4°
43.7°	0.6909	0.7230	0.9556	1.0464	1.3832	1.4474	46.3°
43.8°	0.6921	0.7218	0.9590	1.0428	1.3855	1.4448	46.2°
43.9°	0.6934	0.7206	0.9623	1.0392	1.3878	1.4422	46.1°
44.0°	0.6947	0.7193	0.9657	1.0355	1.3902	1.4396	46.0°
44.1°	0.6959	0.7181	0.9691	1.0319	1.3925	1.4370	45.9°
44.2°	0.6972	0.7169	0.9725	1.0283	1.3949	1.4344	45.8°
44.3°	0.6984	0.7157	0.9759	1.0247	1.3972	1.4318	45.7°
44.4°	0.6997	0.7145	0.9793	1.0212	1.3996	1.4293	45.6°
	cos A	sin A	cot A	tan A	csc A	sec A	A

A	sin A	cos A	tan A	cot A	sec A	csc A	
44.5°	0.7009	0.7133	0.9827	1.0176	1.4020	1.4267	45.5°
44.6°	0.7022	0.7120	0.9861	1.0141	1.4044	1.4242	45.4°
44.7°	0.7034	0.7108	0.9896	1.0105	1.4069	1.4217	45.3°
44.8°	0.7046	0.7096	0.9930	1.0070	1.4093	1.4192	45.2°
44.9°	0.7059	0.7083	0.9965	1.0035	1.4117	1.4167	45.1°
45.0°	0.7071	0.7071	1.0000	1.0000	1.4142	1.4142	45.0°
	cos A	sin A	cot A	tan A	csc A	sec A	A

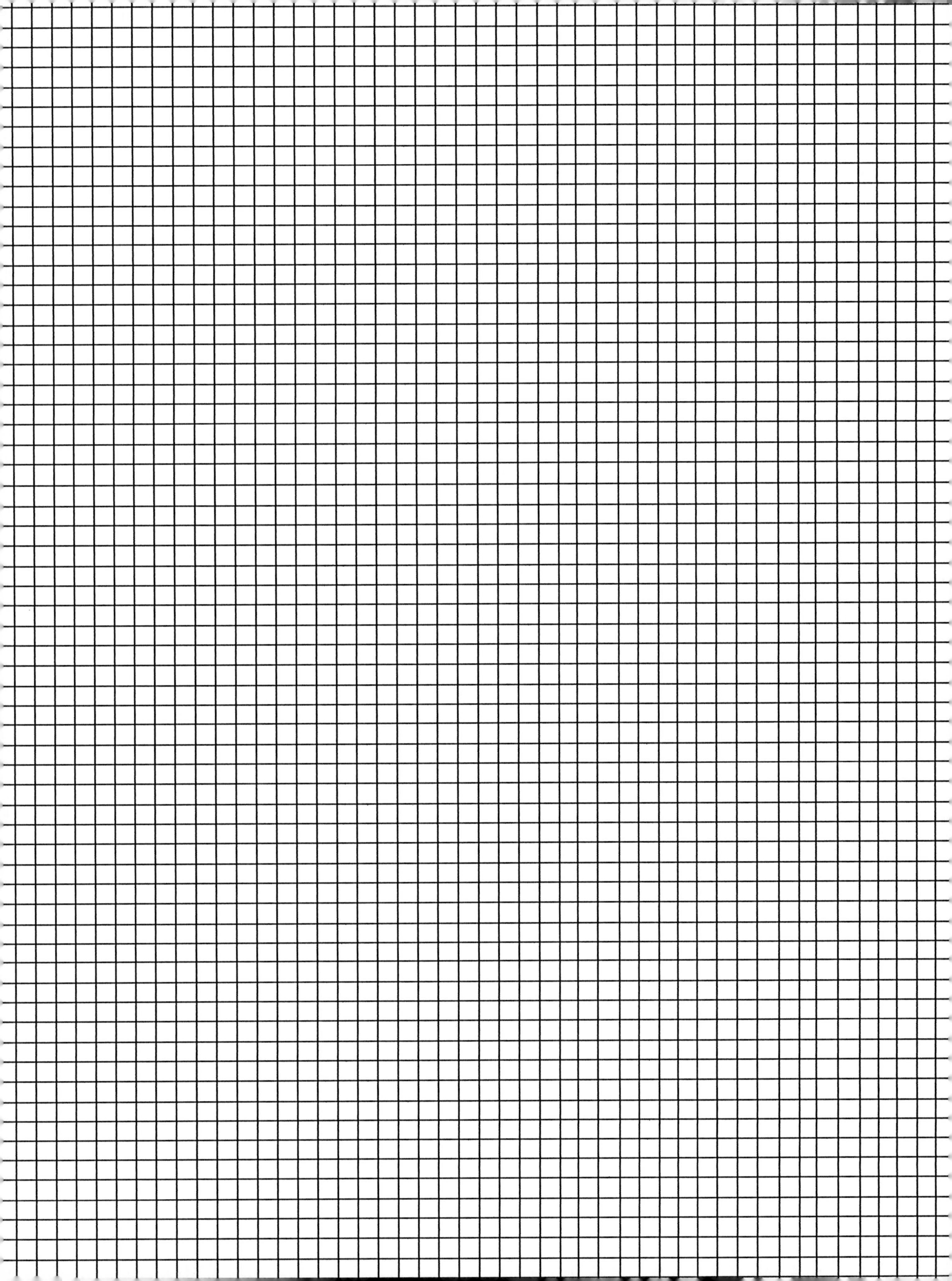

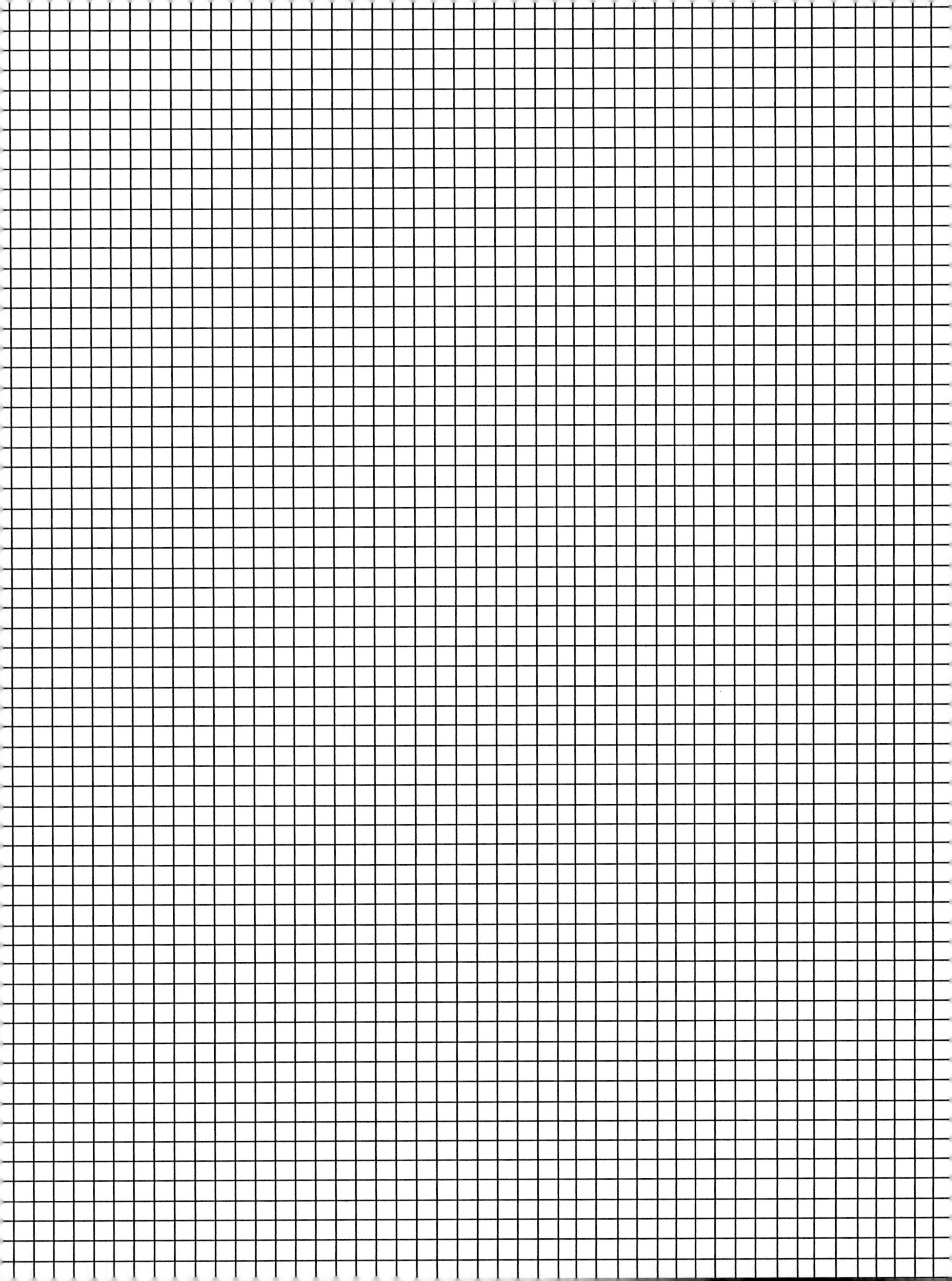

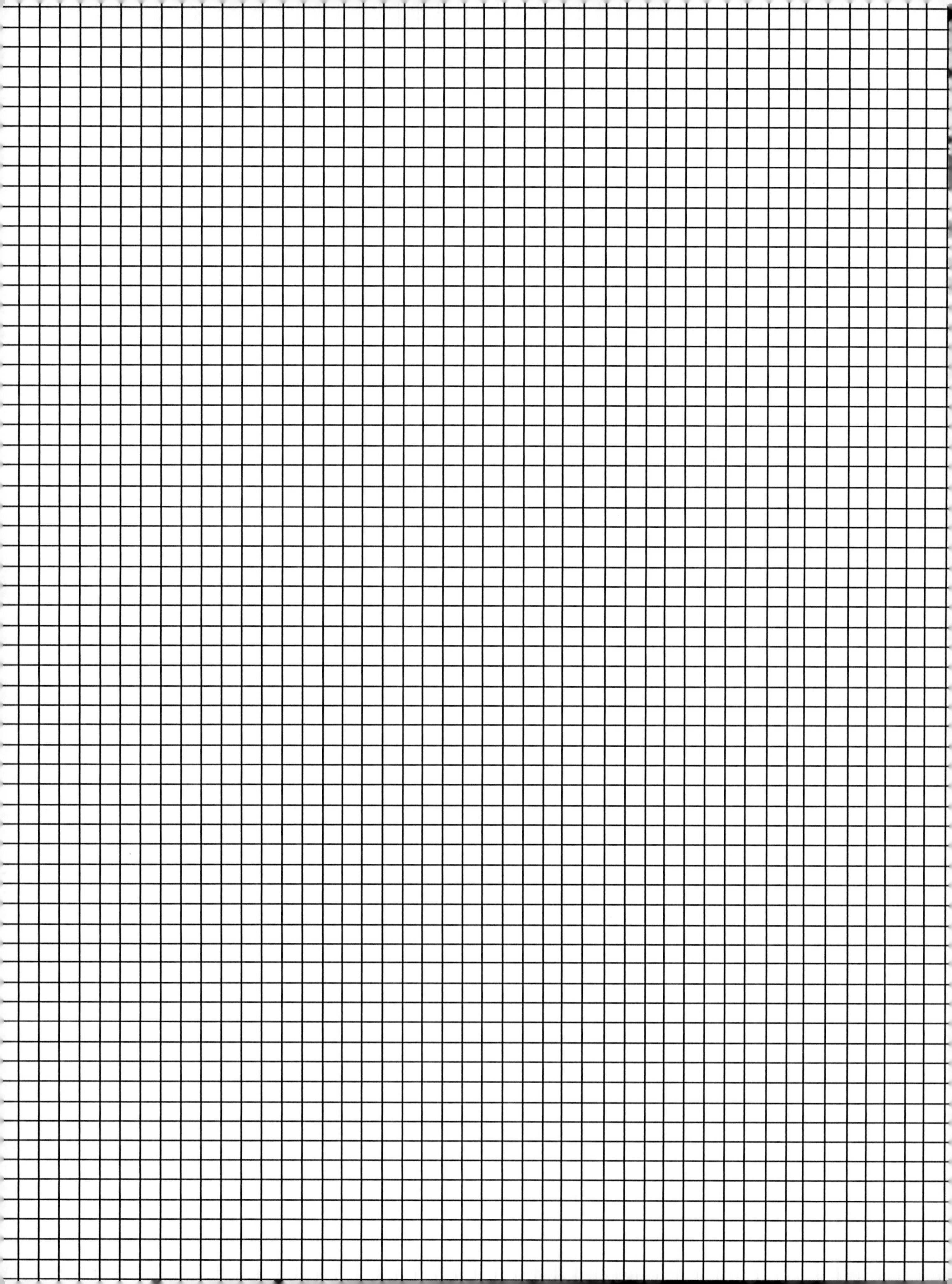

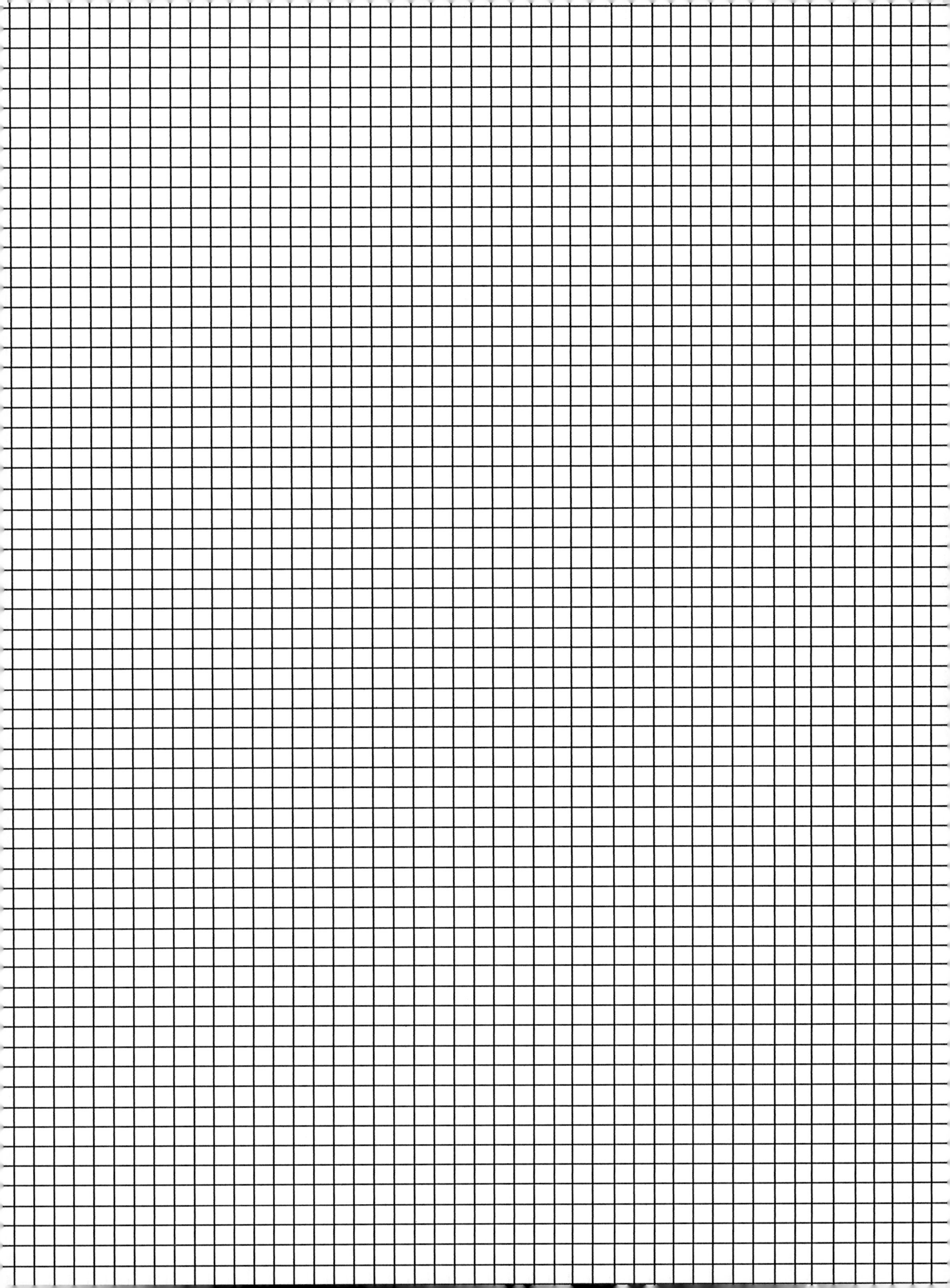

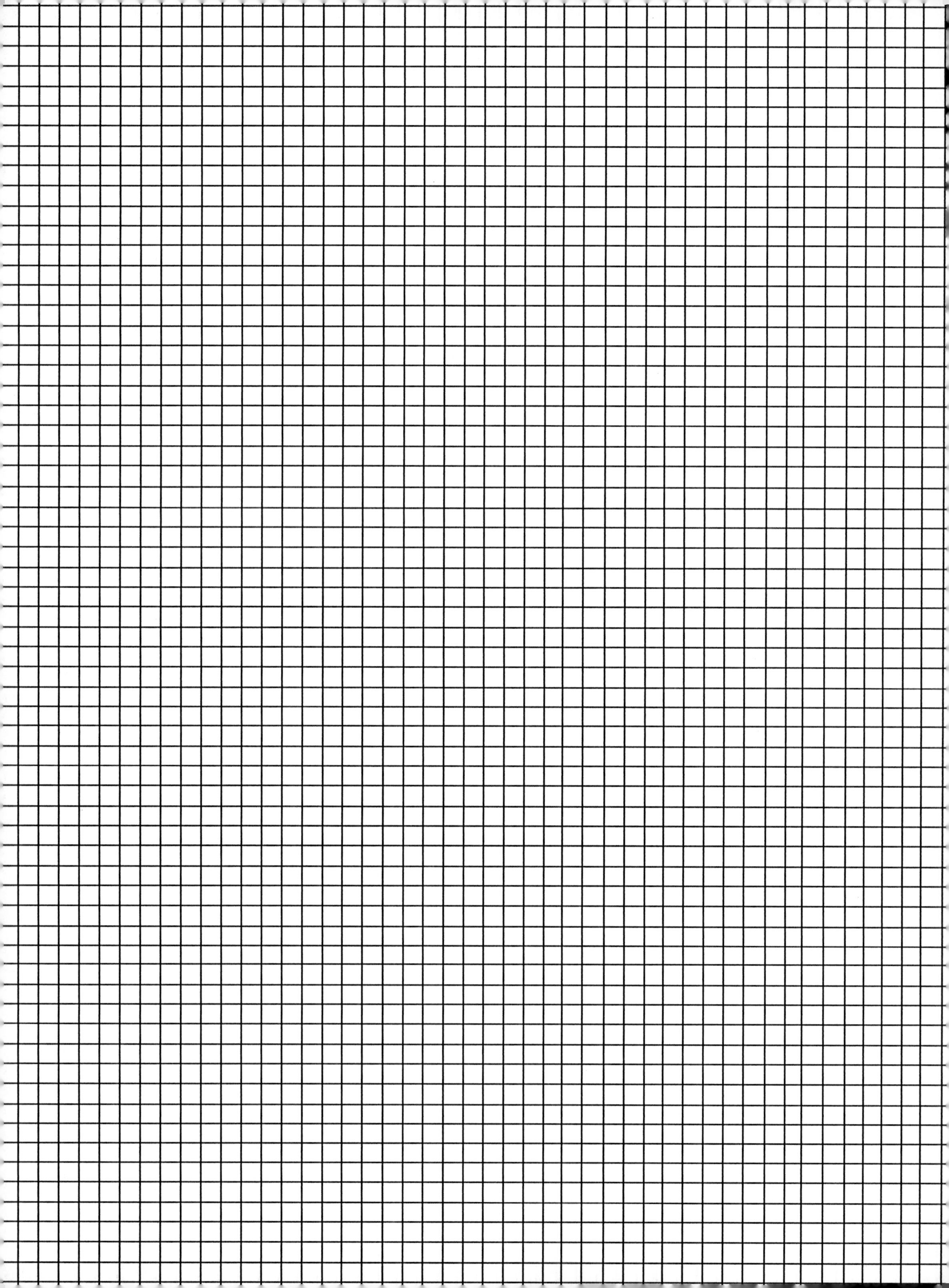

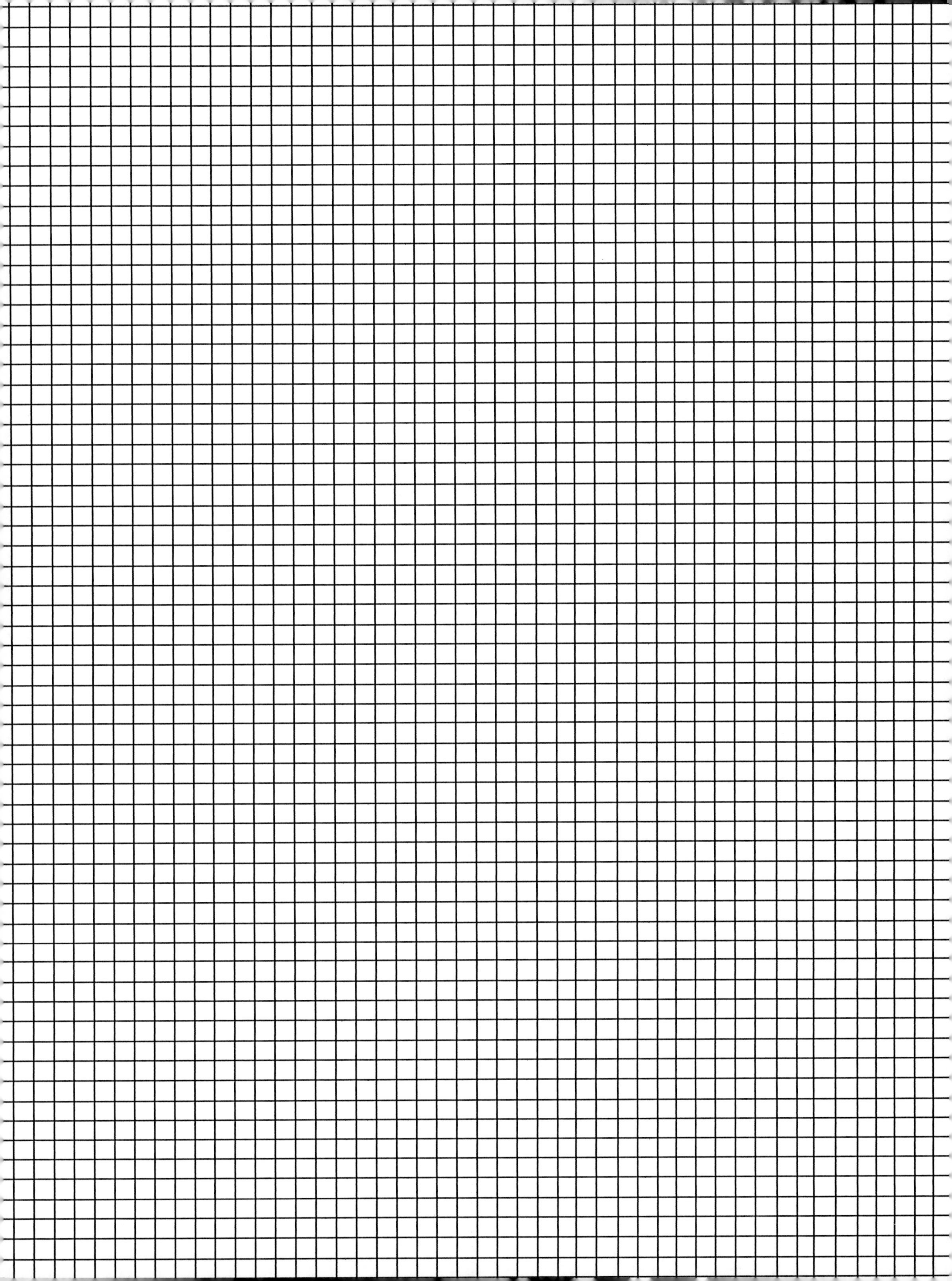

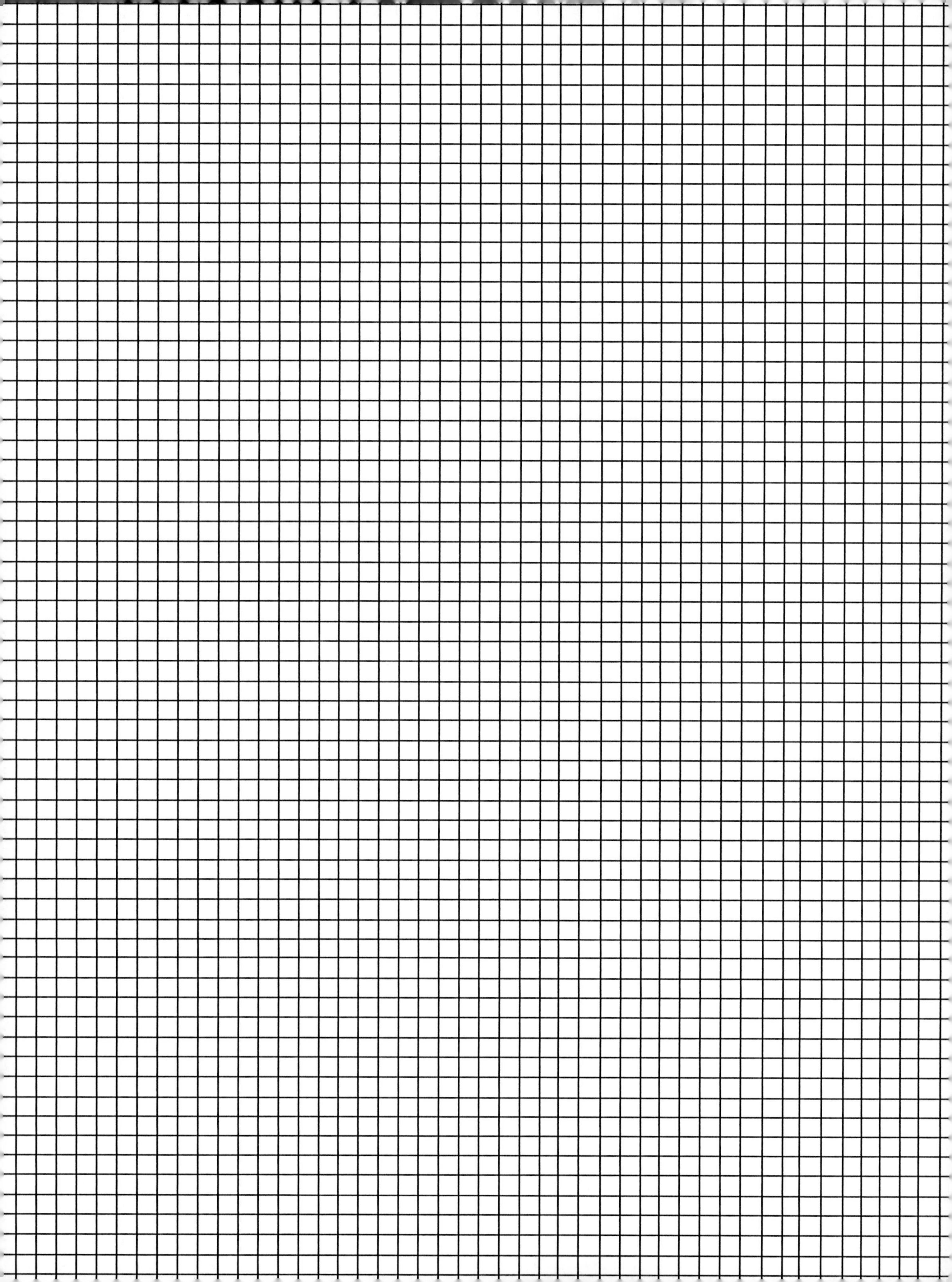

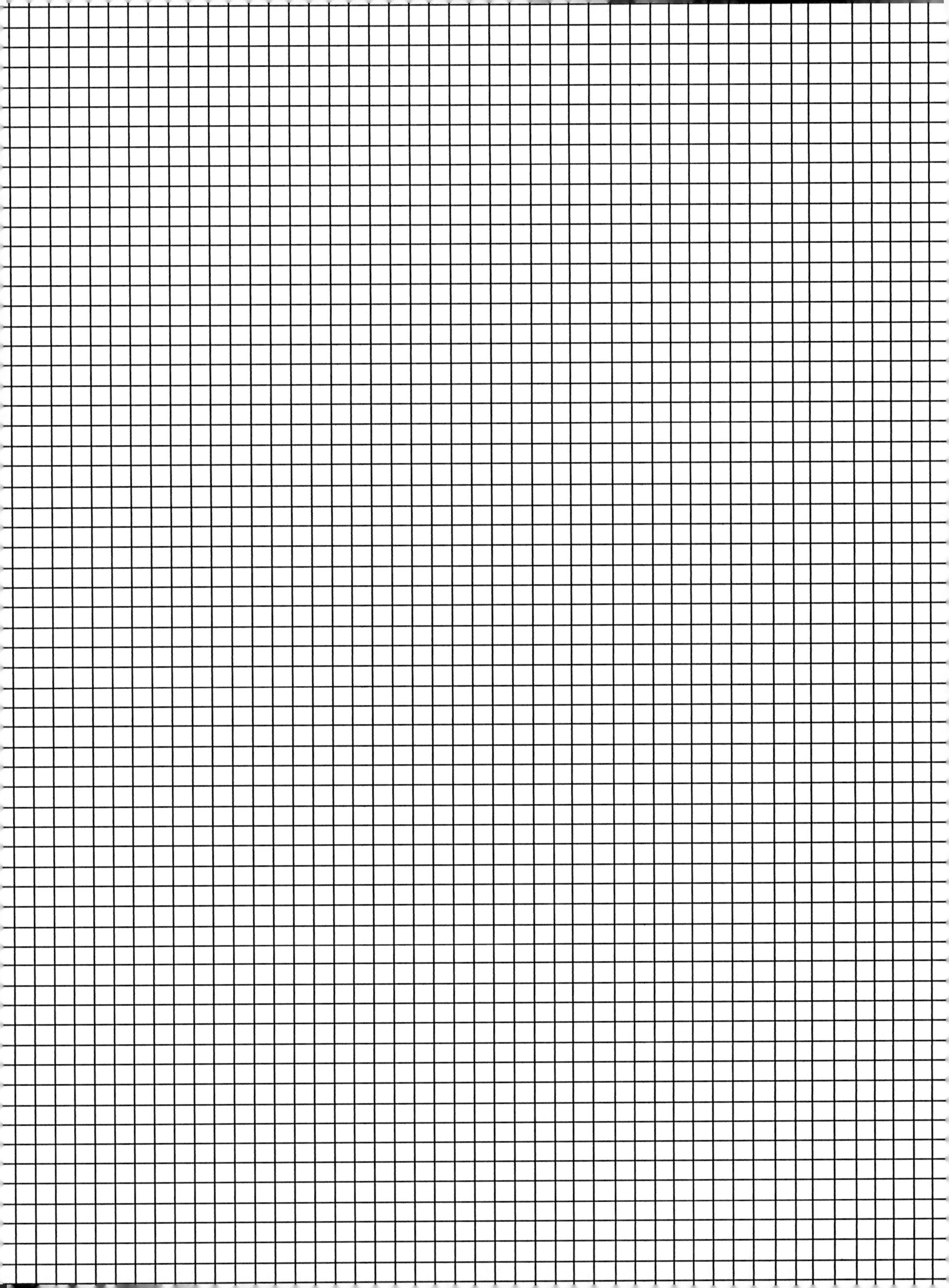

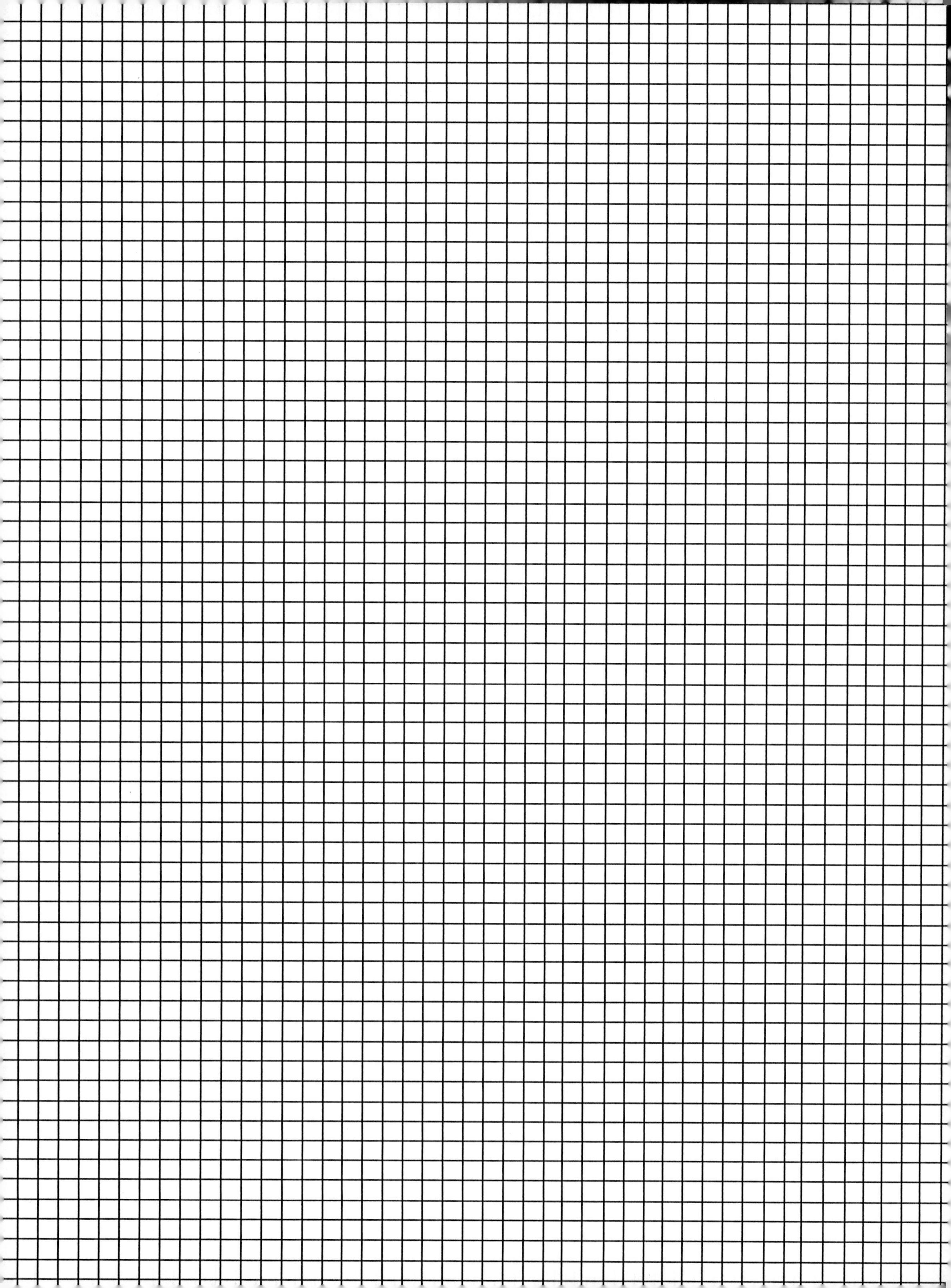

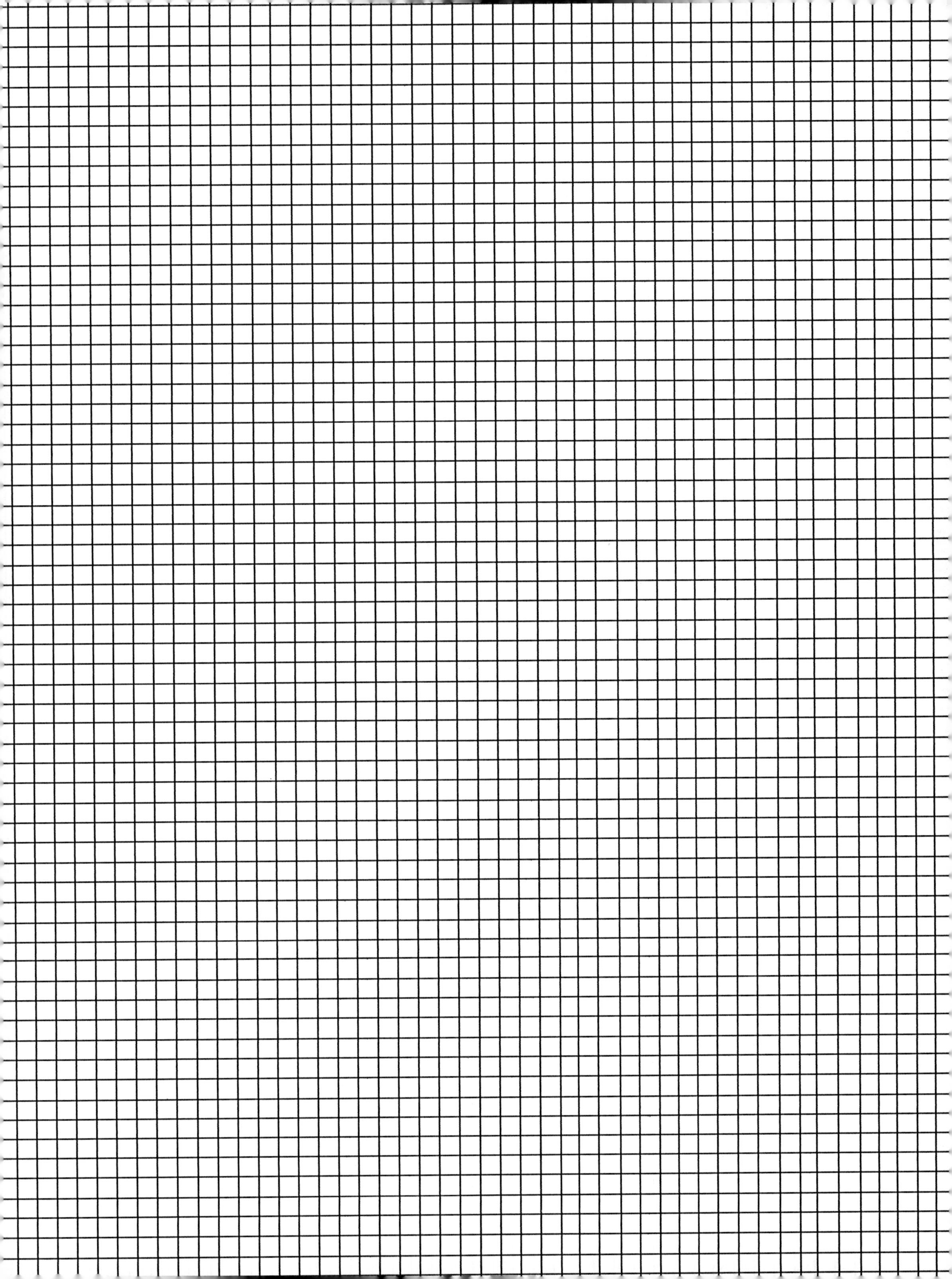

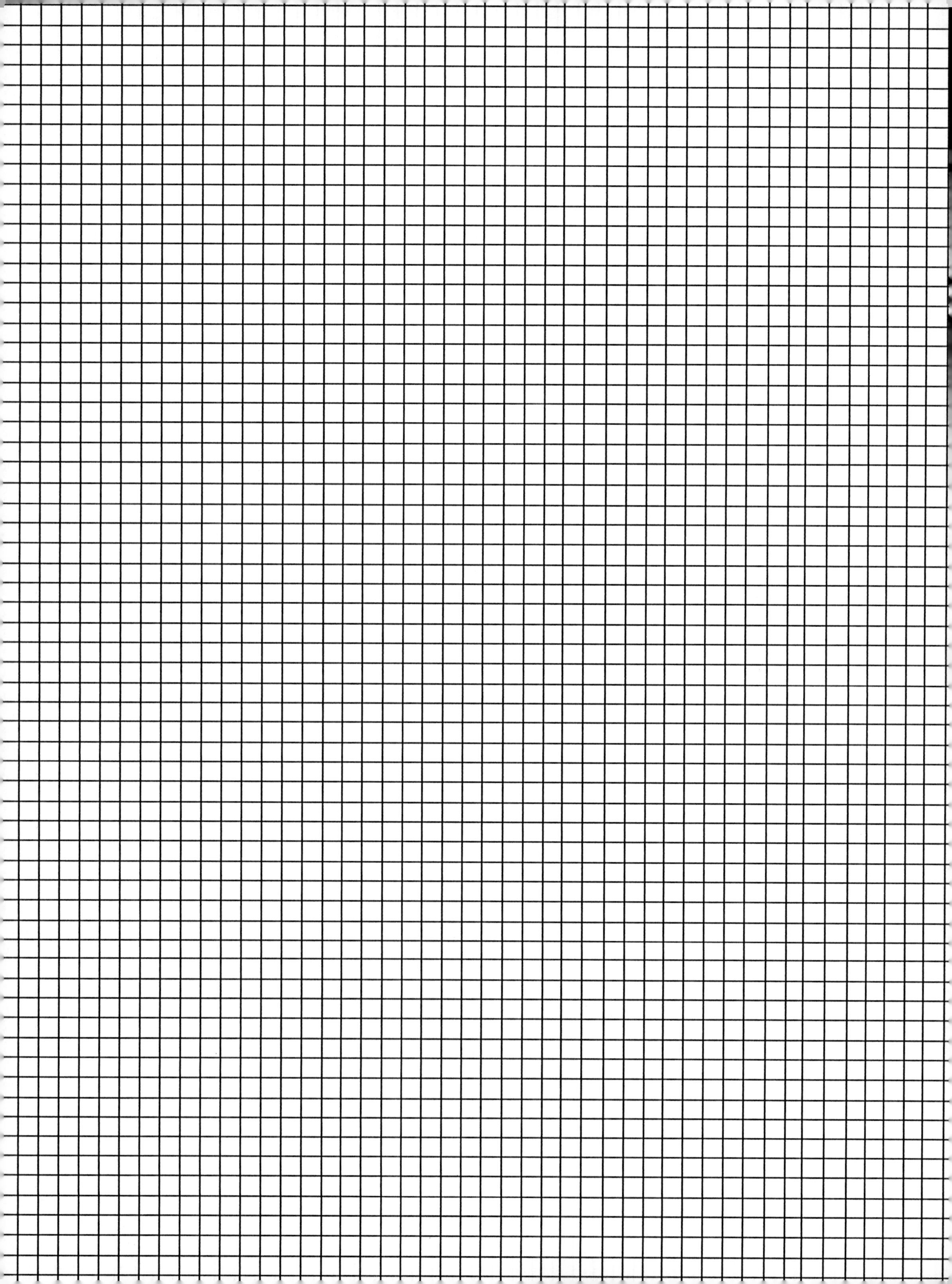

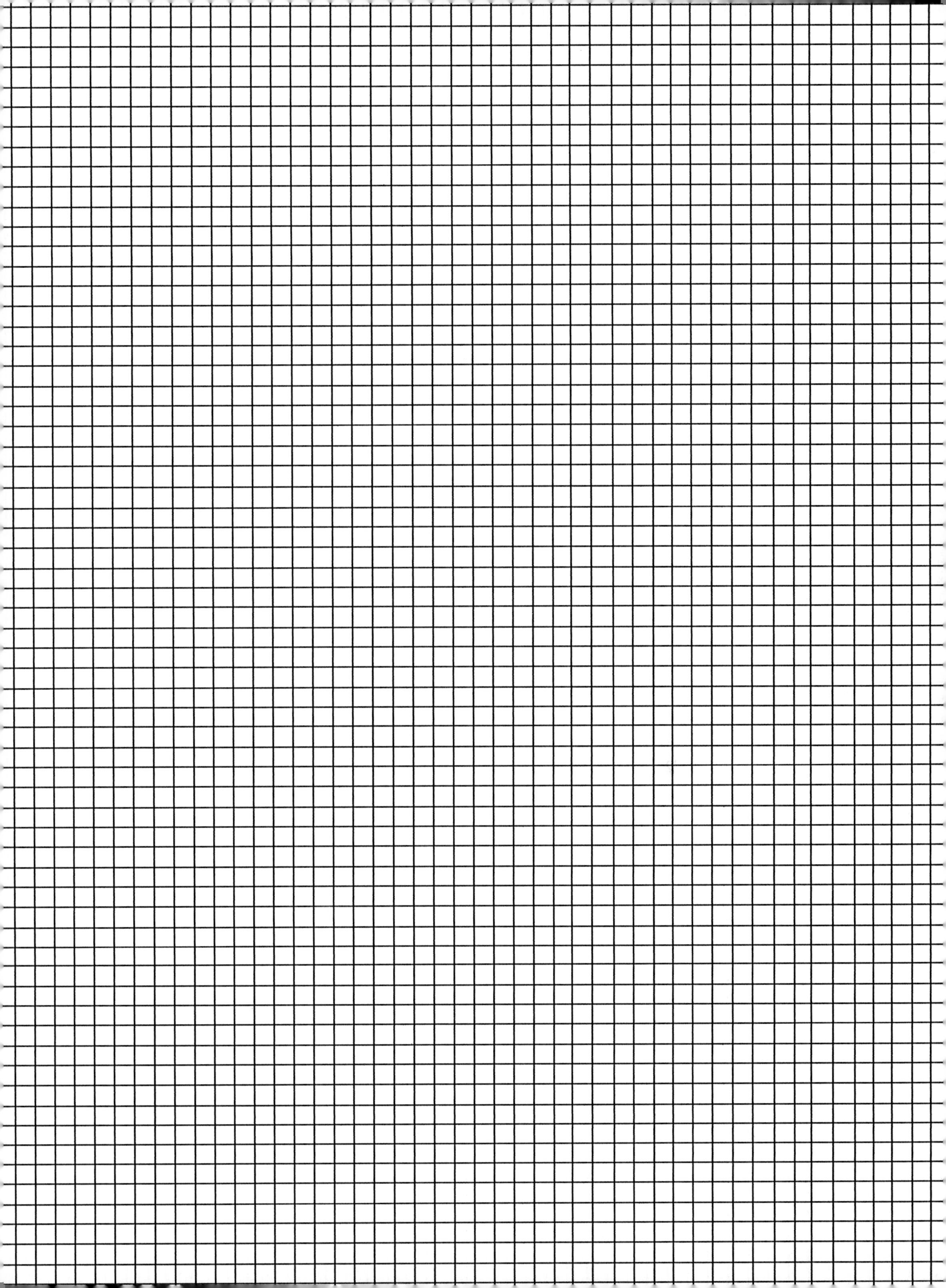

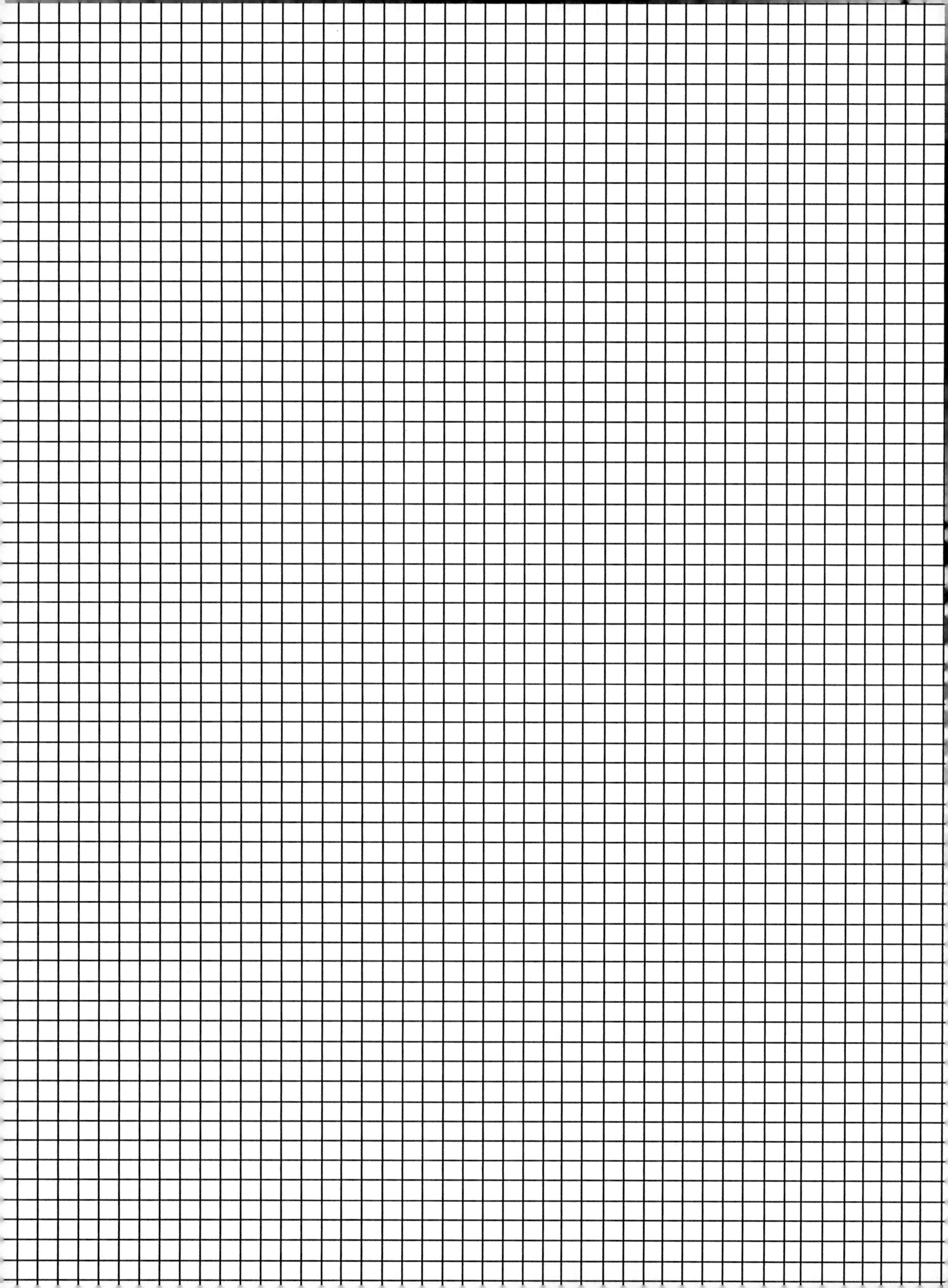

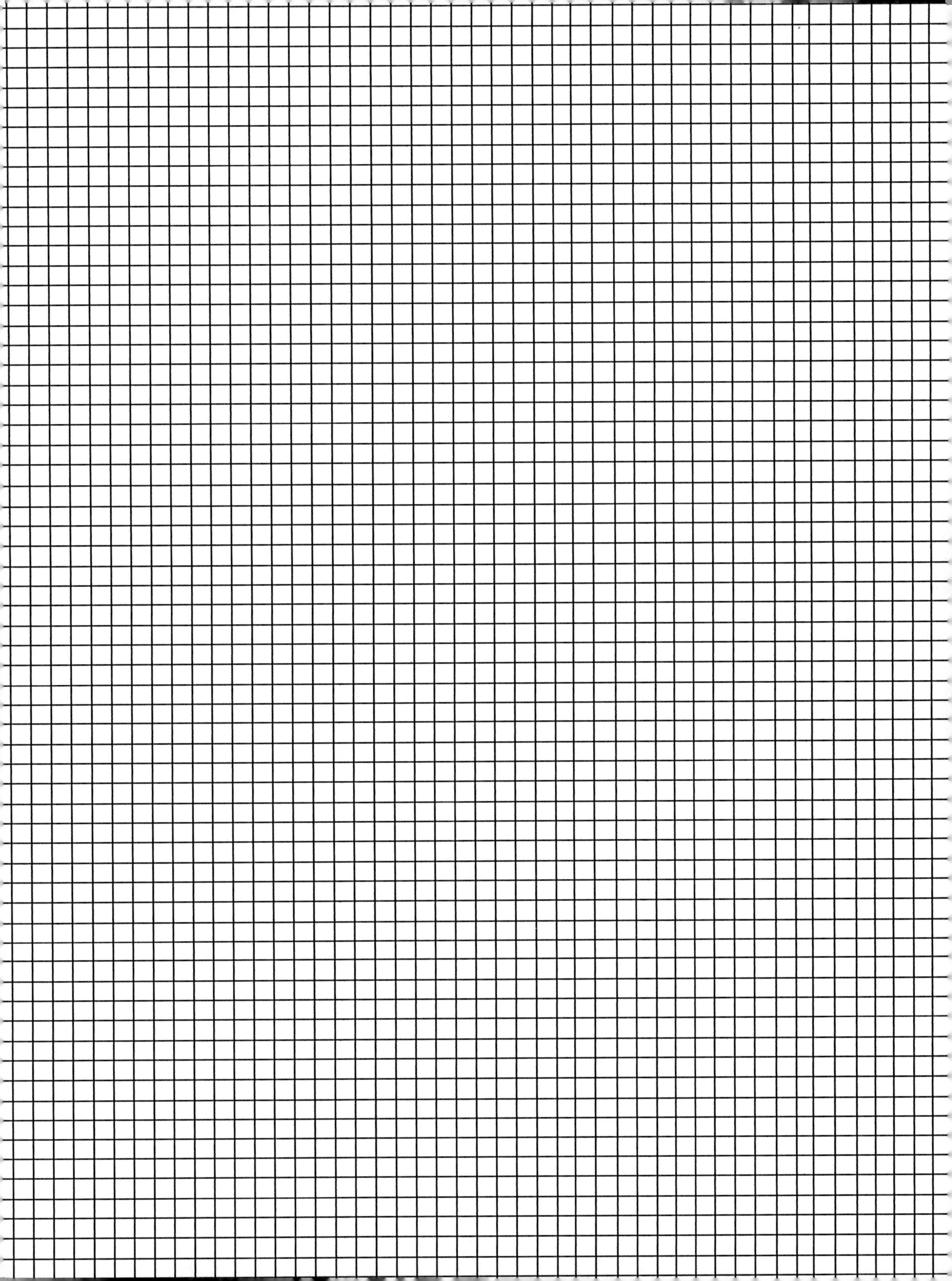

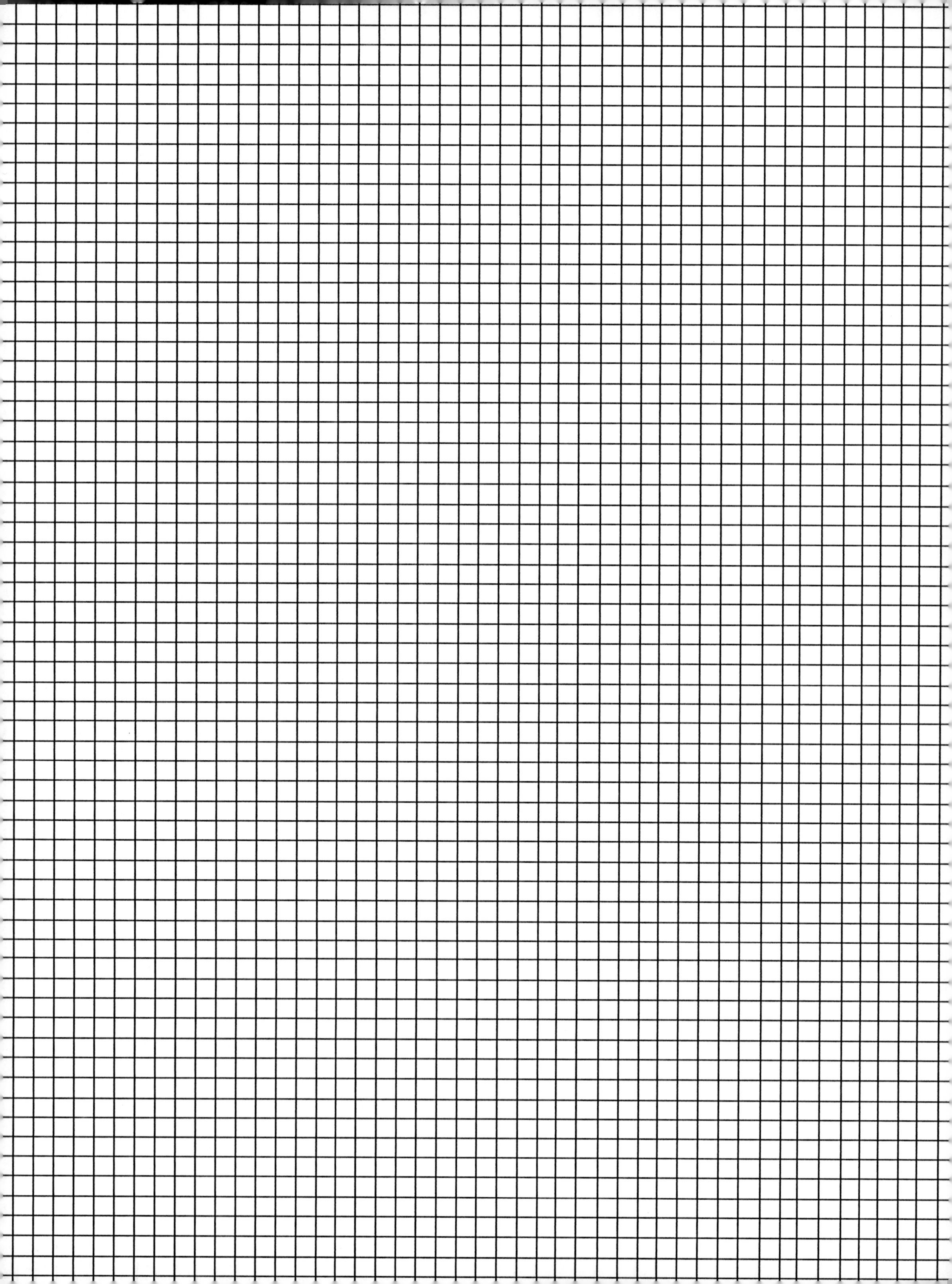

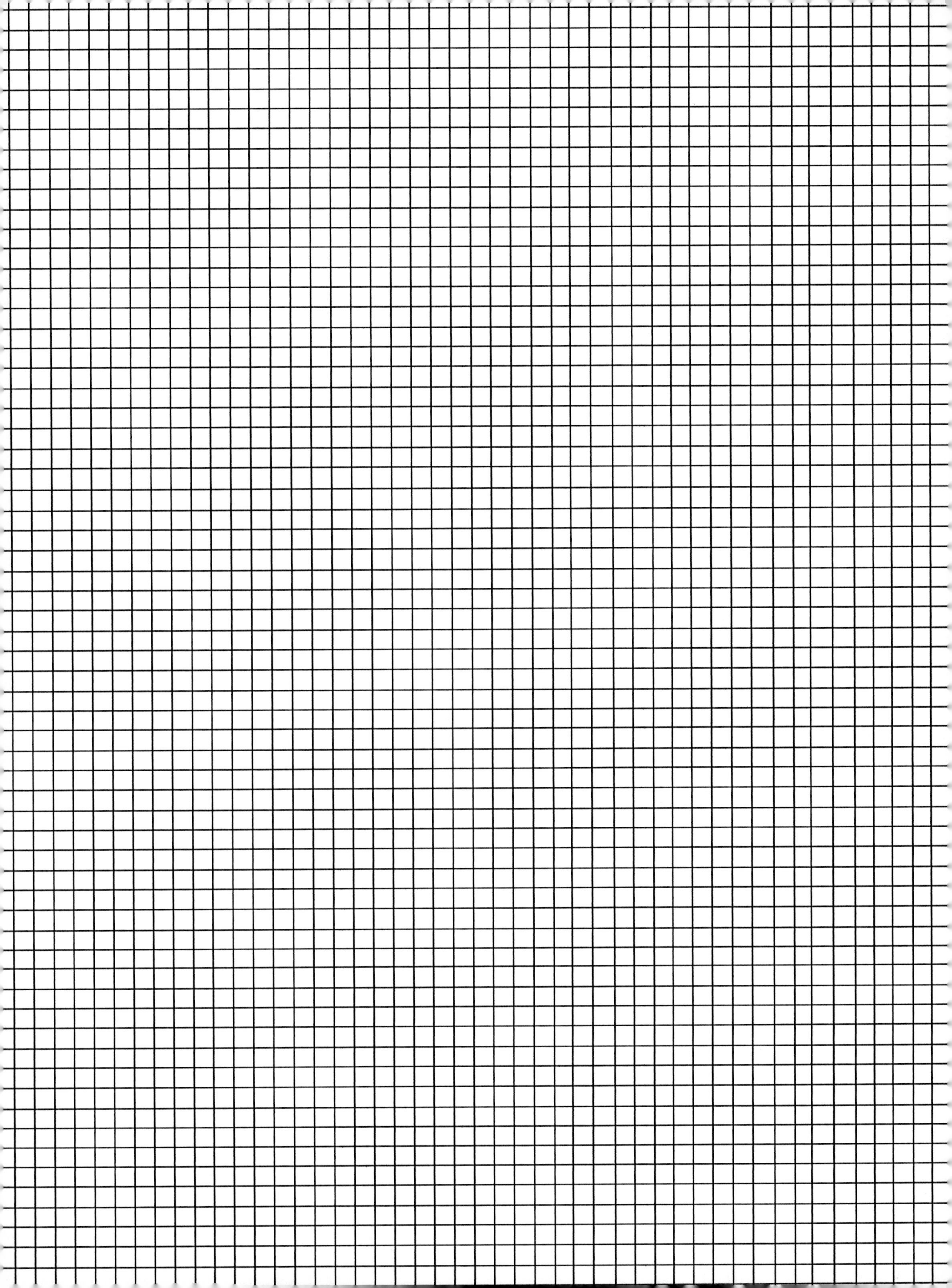

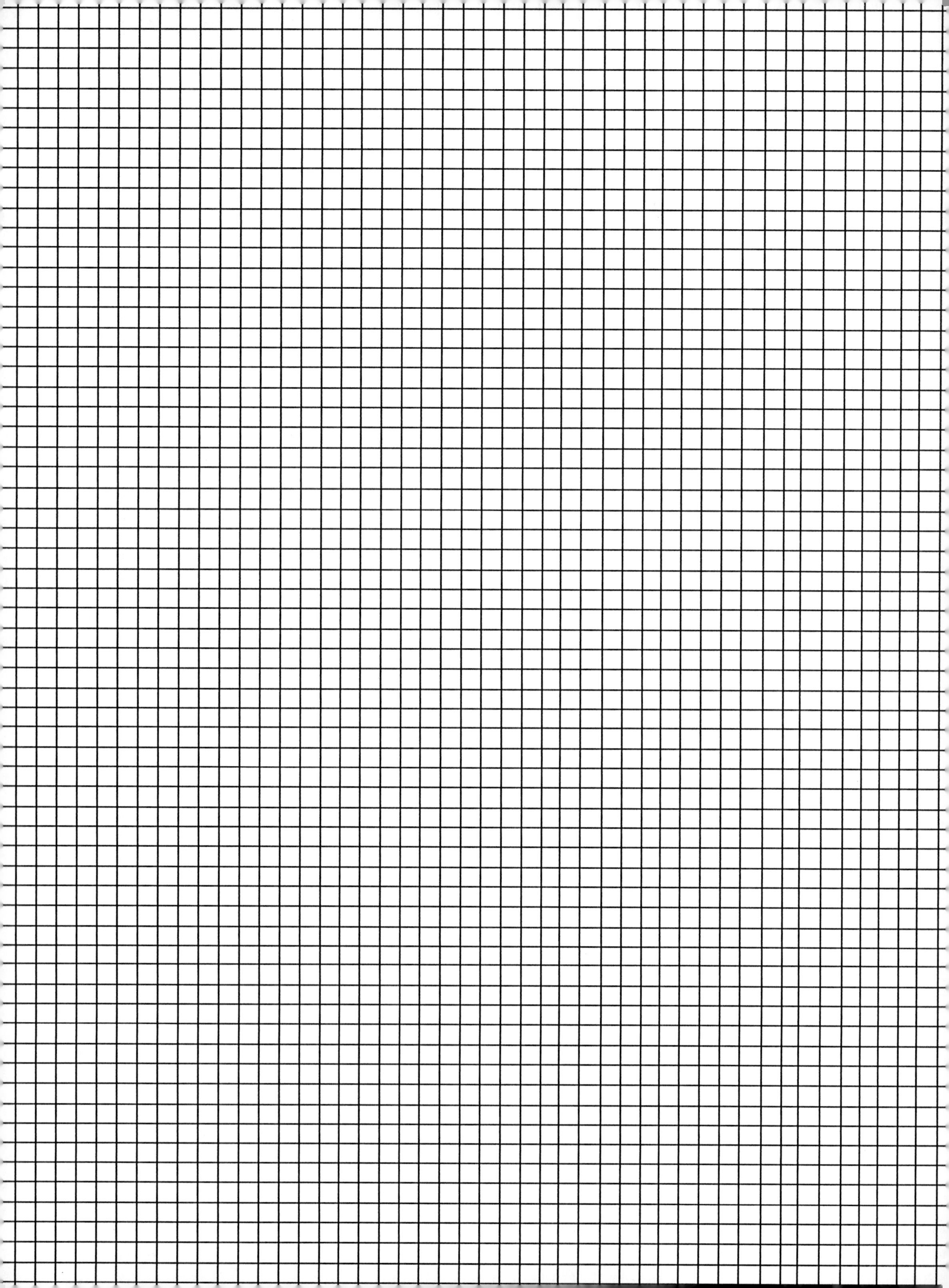

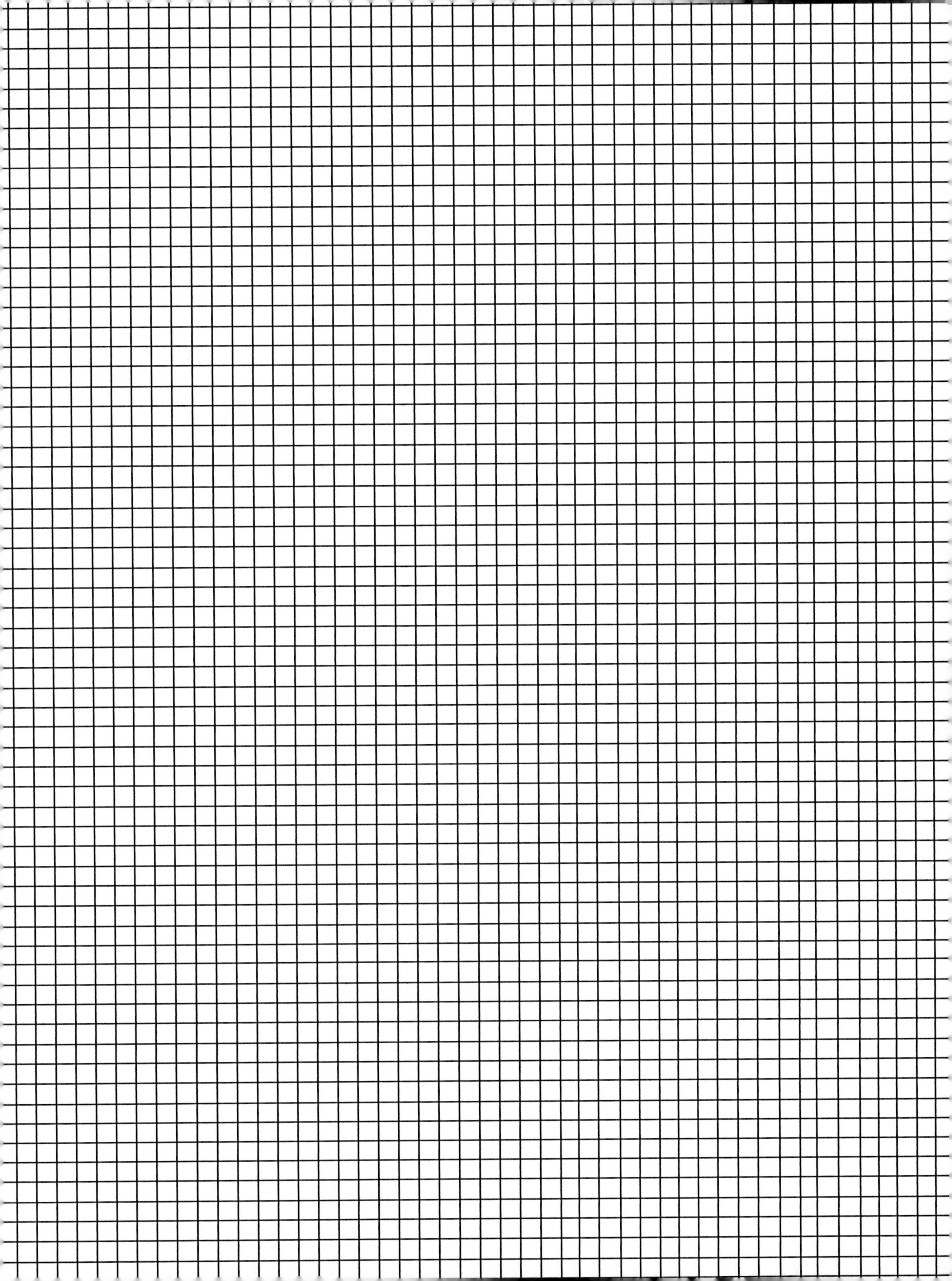

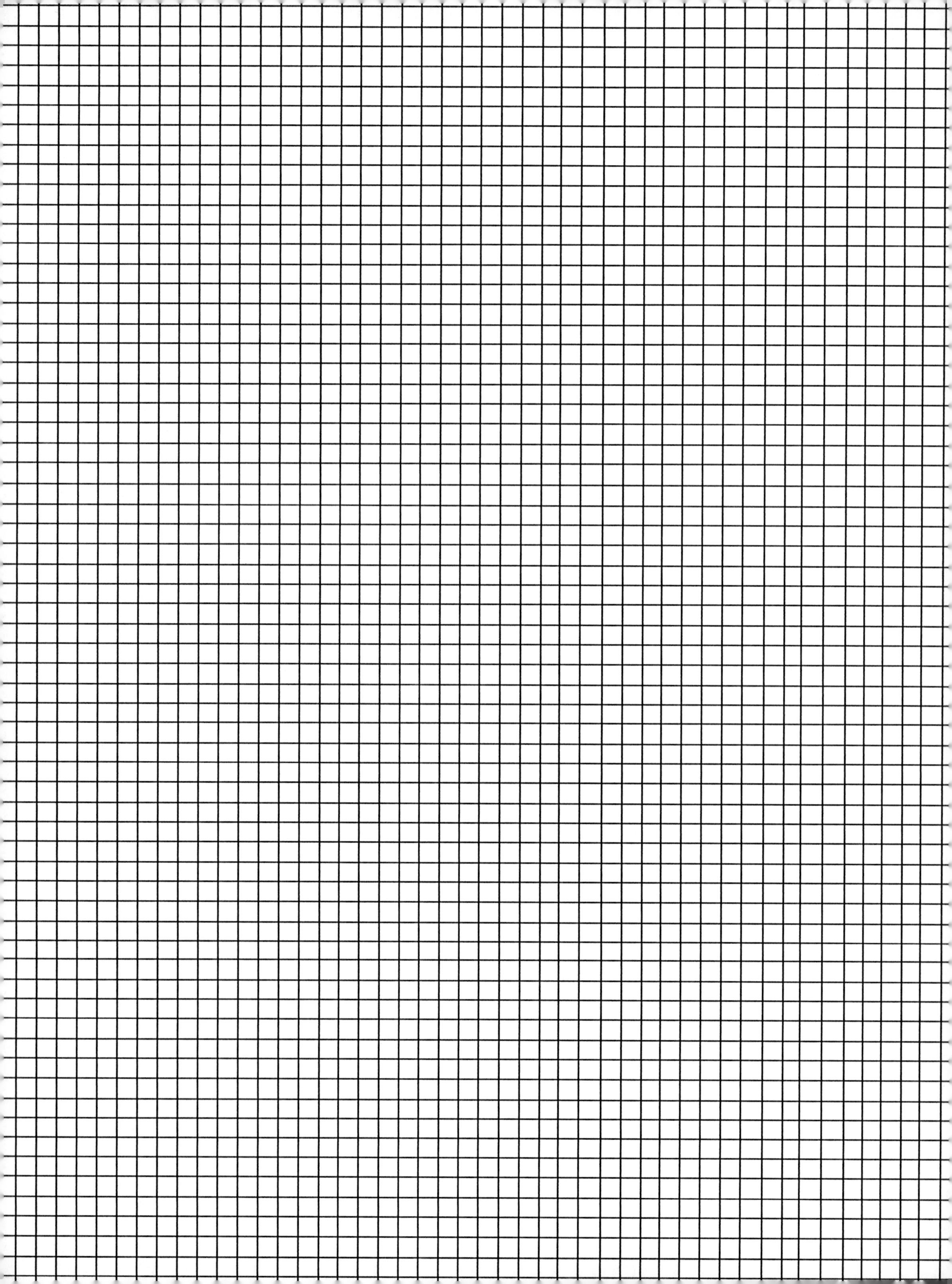

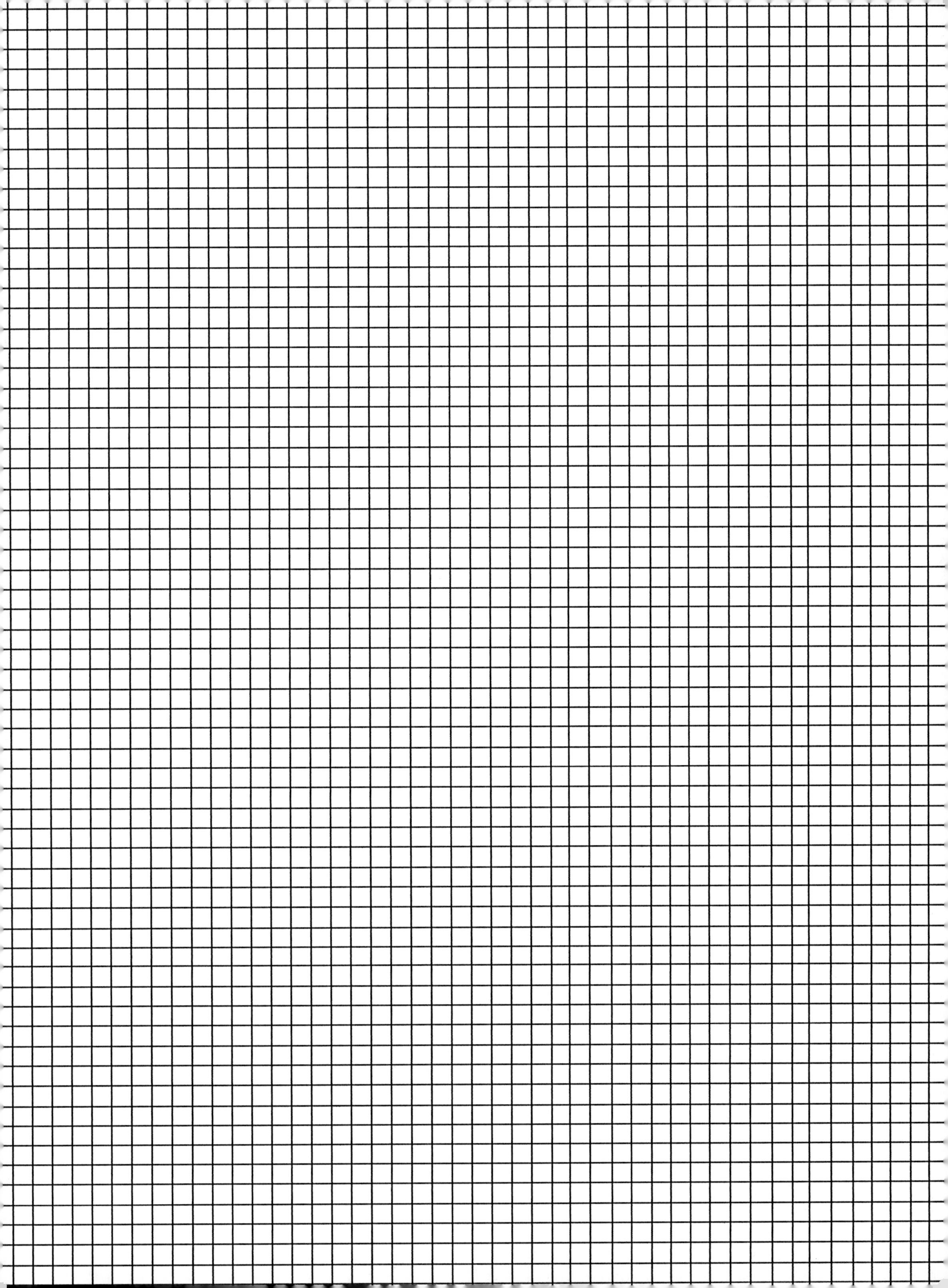

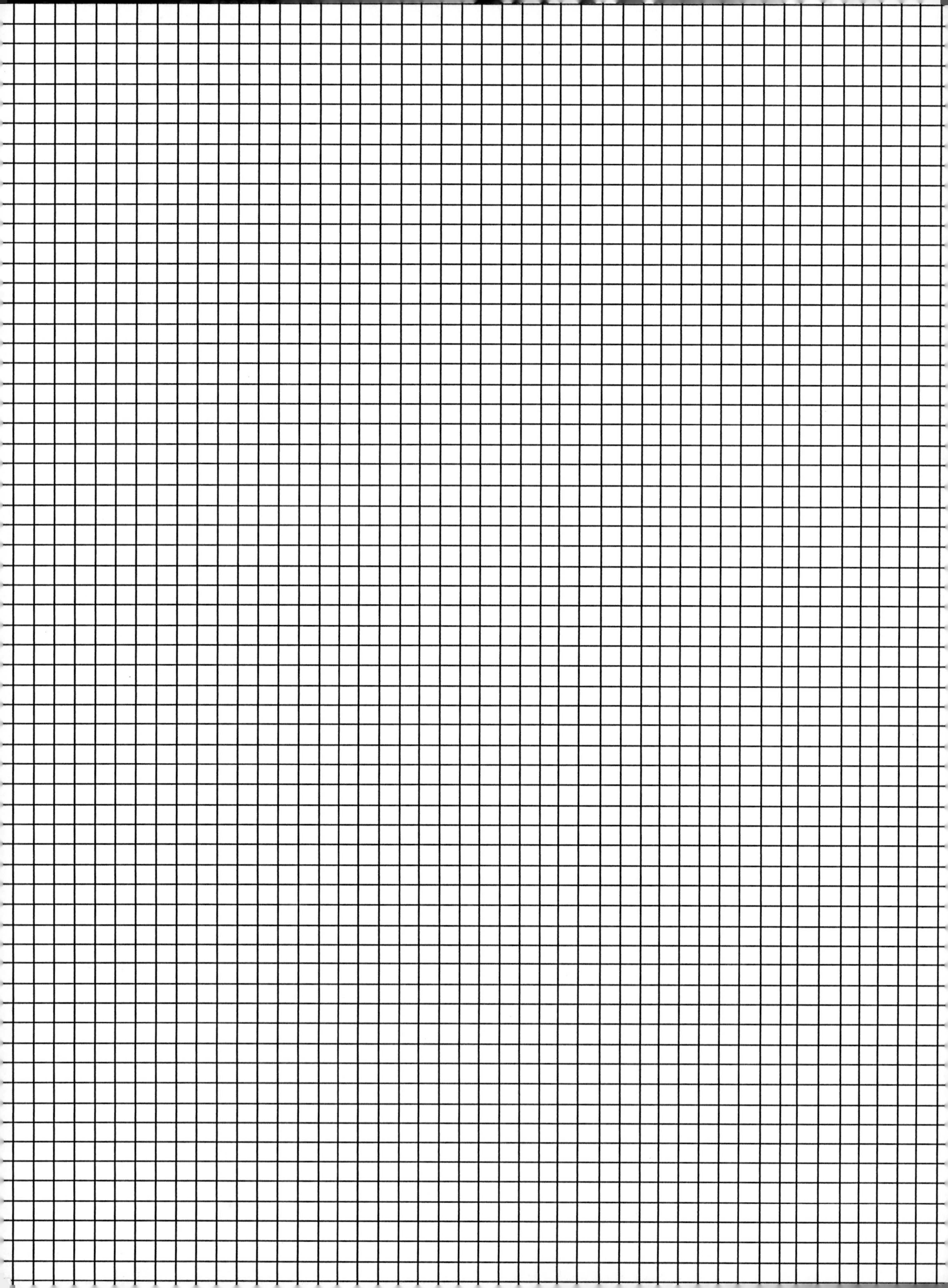

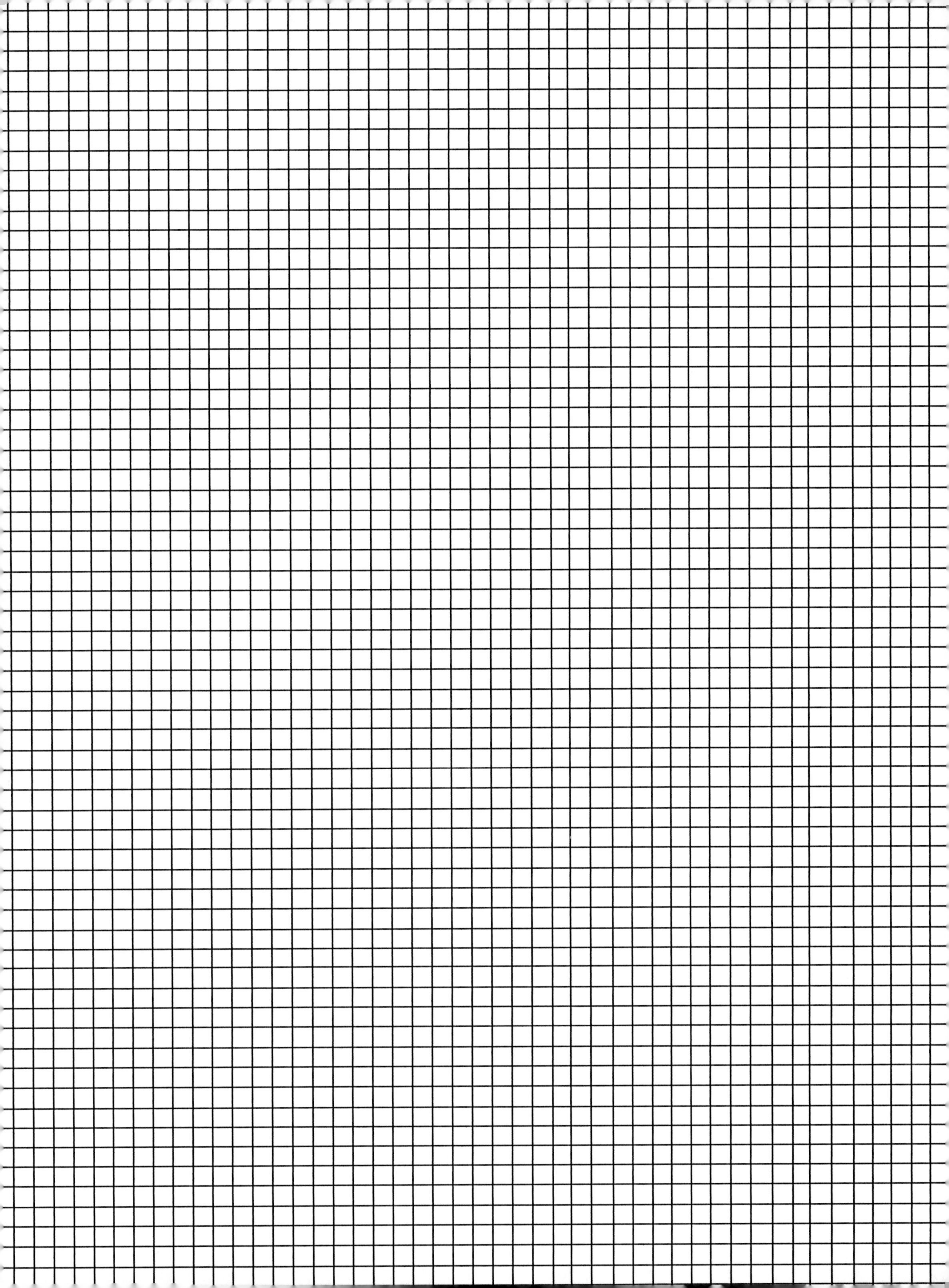

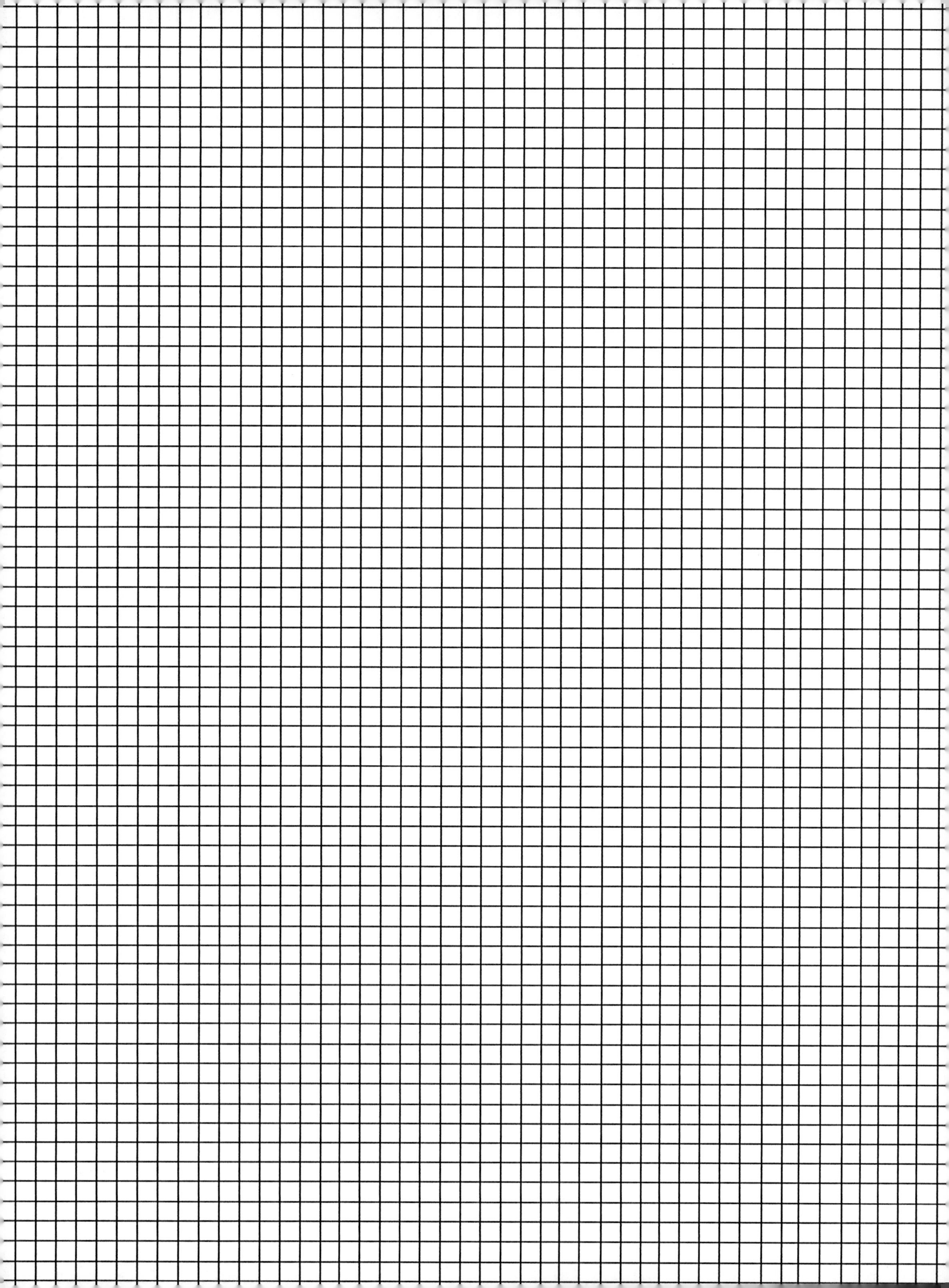

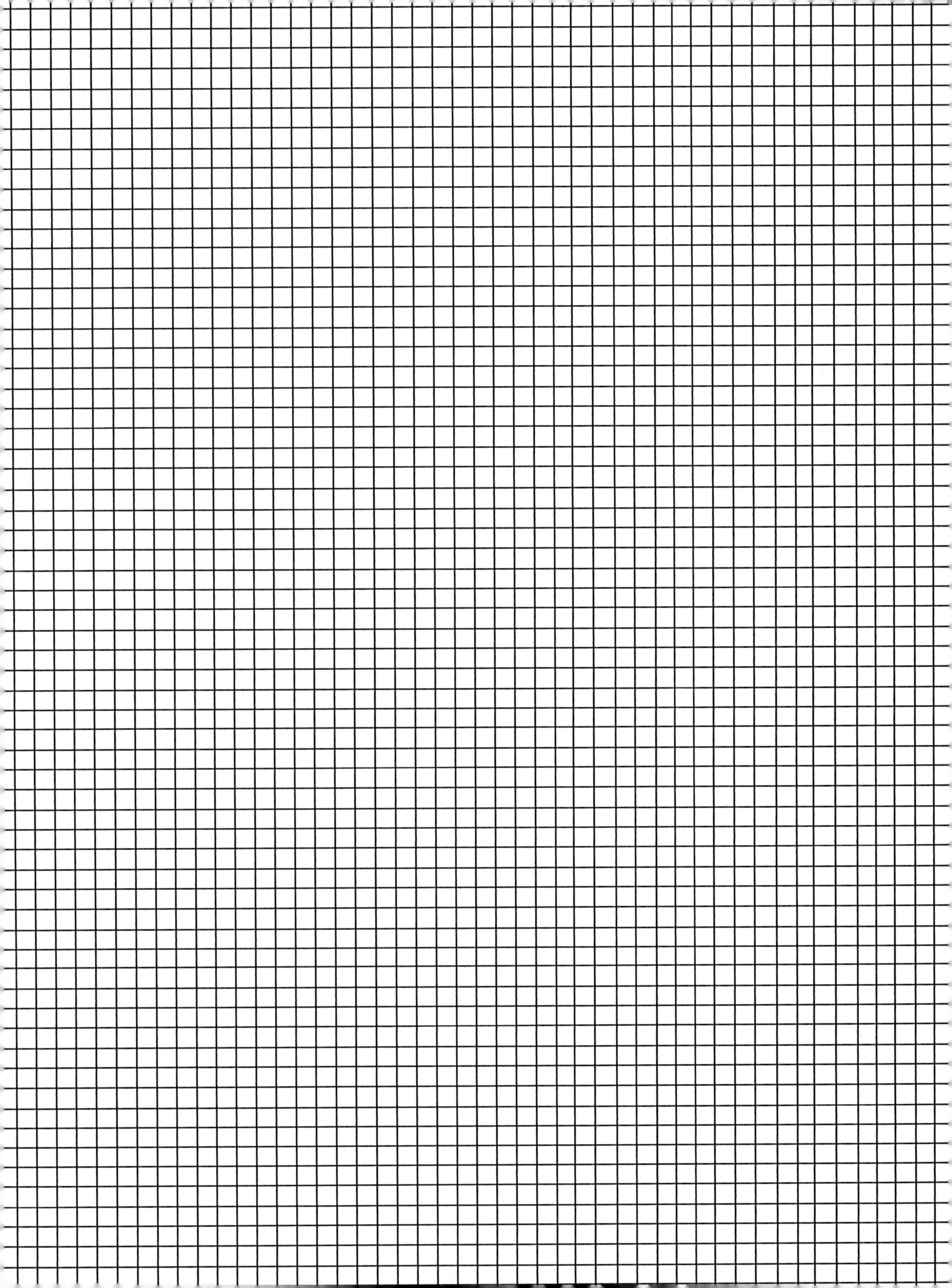